Procedures to Investigate

Procedures to Investigate Foodborne Illness

Sixth Edition

Prepared by the
Committee on the Control of Foodborne Illness
International Association for Food Protection

ISBN 978-1-4419-8395-4   e-ISBN 978-1-4419-8396-1
DOI 10.1007/978-1-4419-8396-1
Springer New York Dordrecht Heidelberg London

Library of Congress Control Number: 2011924223

Printed on acid-free paper

Springer is part of Springer Science+Business Media (www.springer.com)

Procedures to Investigate Foodborne Illness

Sixth Edition

Prepared by the
Committee on the Control of Foodborne Illness International Association
for Food Protection

The following Committee members contributed to this Edition

Ewen C. D. Todd, PhD
Committee Chair
President, Ewen Todd Consulting
Okemos, MI USA

Charles A. Bartleson, M.P.H., R.S.
Washington State Department of Health (retired)
Olympia, WA USA

John J. Guzewich, M.P.H., R.S.
Center for Food Safety and Applied Nutrition
Food and Drug Administration
College Park, MD USA

Agnes Tan, B. Tech (Food) (Hons), M. Hlth. Admin
Microbiological Diagnostic Unit, Public Health Laboratory
Department of Microbiology & Immunology
University of Melbourne
Melbourne
Australia

Marilyn Lee, M.Sc., CPHI(C)
School of Occupational and Public Health
Ryerson University
Toronto, Ontario, Canada

Maria Nazarowec-White, PhD
Associate Director
Bureau of Microbial Hazards
Food Directorate, Health Products and Food Branch
Health Canada
Ottawa, Canada

Foreword

This sixth edition of *Procedures to Investigate Foodborne Illness* is designed to guide public health personnel or teams in any country that investigates reports of alleged foodborne illnesses. It is a companion to another manual published by the IAFP, *Procedures to Investigate Waterborne Illness*. These two manuals are based on epidemiologic principles and investigative techniques that have been found effective in determining causal factors of disease incidence.

This edition describes procedures to:

- Plan, prepare, investigate, and respond to unintentional and intentional contamination of food
- Handle illness alerts and food-related complaints that may be related to illness
- Interview ill persons, those at risk, and controls
- Develop a case definition
- Collect and ship specimens and samples
- Conduct hazard analysis at sites where foods responsible for outbreaks were produced, processed, or prepared
- Trace sources of contamination
- Identify factors responsible for contamination, survival of pathogenic microorganisms or toxic substances, and/or propagation of pathogens
- Collate and interpret collected data
- Report information about the outbreak

These guidelines are presented in the sequence usually followed during investigations. They are organized so that an investigator can easily find the information needed in any phase of an investigation. The instructions, forms, tables, keys, and information on selected etiologic agents, diseases, and food vehicles are comprehensive but not exhaustive.

This manual combines the principles and techniques of epidemiology, statistics, and food preparation review that guide the formation of rational hypotheses and their testing. The *Table of Contents* serves as an outline, and a flow diagram shows interrelationships of the activities and their typical sequence in an investigation. This edition, which includes a section on intentional contamination of food, has a completely revised set of keys in color, and the format of the Manual is larger size for better ease of reading. There will also be a complementary electronic version.

The final foodborne illness summary report form contains detailed categories of places foods where acquired; the sites of contamination, survival, and propagation, which may be at different locations; method of processing or preparation; factors that contributed to the outbreak; and space for a narrative summary. Collection of these sorts of data, over time, can provide focus for preventive and control measures.

Kits including the manual, expanded-size forms, and equipment useful for investigation can be preassembled so that upon notification of an outbreak, investigation can be initiated without delay.

This manual supersedes previous editions by providing updated information to assist in conducting thorough investigations of foodborne disease outbreaks.

Contents

Procedures to Investigate Foodborne Illness

Introduction

During production, harvesting, processing, packaging, transportation, preparation, storage, and service, any food may be exposed to contamination with poisonous substances or infectious or toxigenic microorganisms. Processing or preparation failure may lead to survival of microorganisms or toxins, and time–temperature abuse can allow proliferation of pathogenic bacteria and molds. In addition, some plants are intrinsically toxic. Animals may acquire toxins from their food or metabolize them, or they become infected with or colonized by pathogenic bacteria, viruses, and parasites. If a product contaminated with sufficient quantities of poisonous substances or pathogenic microorganisms is eaten, susceptible persons may develop foodborne illness. The food supply may be local or global. In fact today, fresh as well as shelf stable food is available from sources all over the world. Foodborne illness outbreaks are routinely being linked to sources of contamination far distant from the point of consumption. Whatever the source, an investigation always begins at one or more local levels and can expand from there. Therefore, foodborne illness surveillance, investigation, and response systems require the close collaboration and coordination of food safety and public health agencies at local, state/provincial, federal/national, and international levels.

Foodborne illnesses comprise the various acute syndromes that result from ingestion of contaminated foods. They are classified as:

- Intoxications caused by ingestion of foods containing either poisonous chemicals or toxins produced by microorganisms
- Toxin-mediated infections caused by bacteria that produce enterotoxins (toxins that affect water, glucose, and electrolyte transfer) during their colonization and growth in the intestinal tract
- Infections caused when microorganisms invade and multiply in the intestinal mucosa or other tissues

Manifestations range from slight discomfort to acute illness to severe reactions that may terminate in death or chronic sequelae, depending on the nature of the causative agent, number of pathogenic microorganisms or concentration of poisonous substances ingested, and host susceptibility and reaction.

Food Protection International Association, *Procedures to Investigate Foodborne Illness*, DOI 10.1007/978-1-4419-8396-1_1, © International Association for Food Protection 2011

The public relies on the food industry and health and food regulatory officials for protection from foodborne illness. Such protection depends on rapid detection of outbreaks and a thorough knowledge of the agents and factors responsible for foodborne illness. Current food distribution systems are able to move contaminated products throughout a country or to any place in the world within days of processing. In addition, public health and law enforcement agency officials should always be alert to the rare possibility of an intentional contamination of food by disgruntled employees or terrorists. Rapid reporting of local outbreaks to the national and international levels can, therefore, improve the surveillance and control of foodborne disease and detection of intentional contamination.

The purposes of a foodborne illness investigation are to stop the outbreak or prevent further exposures by:

- Identifying illnesses associated with an incident and verifying that the causative agent is foodborne
- Detecting all cases, the causative agent, the implicated food(s), and the place(s) where food was mishandled or mistreated
- Determining the source and mode of contamination, processes, or practices by which proliferation and/or survival of the etiologic agent occurred
- Gathering information on the epidemiology of foodborne diseases and the etiology of the causative agents that can be used for education, training, and program planning, thereby impacting on the prevention of foodborne illness
- Determining if the outbreak under investigation is a part of a larger outbreak by immediately reporting to state/provincial/national epidemiologists

Once the responsible food has been identified, further cases can often be prevented by halting its distribution and sale; recalling production lots already distributed; and quarantining, reprocessing, or disposing of the contaminated foods to prevent their entry or reentry into food channels.

As epidemiologic data accumulate, information will indicate critical control points in food production, processing, and preparation, as well as appropriate methods for controlling and preventing foodborne disease. This information will guide administrators in making rational decisions and setting program priorities to provide the highest degree of food safety at the lowest cost.

A flow chart, *Sequence of events in investigating a typical outbreak of foodborne illness* (Figure A, page 138) shows the sequential steps, as presented in this manual, in investigating a typical outbreak of foodborne illness and indicates their relationships. A description of each step is presented in this manual.

Develop a Foodborne Disease Surveillance System

An effective surveillance system will:

- Systematically collect data pertaining to the occurrence of foodborne illness and outbreaks

- Guide investigations of diseases, clusters of illness, and outbreaks associated with food consumption
- Determine if the outbreak under investigation is part of a larger multijurisdictional outbreak and coordinate that larger investigation
- Enable analysis and interpretations of surveillance and investigational data
- Disseminate consolidated surveillance information to appropriate agencies and public health partners

Many types of disease-detection and disease-reporting systems may already exist at the local or state/provincial level and can be incorporated into a foodborne disease surveillance program. These include:

- Mandatory (or voluntary) laboratory- or physician-based reporting of specific infectious diseases
- Laboratory-based surveillance systems such as PulseNet in the USA
- Medical records from hospital emergency rooms, urgent-care clinics, and physician offices
- Public complaints made to health agencies
- Follow up of complaints of illnesses arising from the potential intentional contamination of food
- School illness and absentee records
- Absentee records of major employers
- Sentinel studies of selected diseases
- Sales of antidiarrheal drugs
- Syndromic surveillance

An effective disease surveillance system is essential for detection of disease caused by either unintentional or intentional contamination of food.

Organize the System and Develop Procedures

An effective foodborne disease surveillance system requires close cooperation between key personnel in public and private health agencies. When your agency contemplates developing or improving a foodborne disease surveillance program, give top priority to obtaining financial, administrative, political, and strategic support. Then, identify a key person to create, implement, and manage the system. This person takes responsibility to:

- Review existing reporting systems that could be incorporated into the foodborne disease surveillance system
- Identify the types of information that cannot be obtained from existing reporting systems but that need to be collected or addressed by the foodborne disease surveillance system
- Identify ways to merge or integrate the data collected by existing systems with data gathered in the foodborne disease surveillance system

- Identify collaborating agencies and staff
- Train agency staff in surveillance methods
- Collaboration with emergency response and law enforcement agency officials when an intentional contamination of food is suspected
- Assemble materials that will be required during an outbreak investigation
- Evaluate the effectiveness of the system

Develop procedures to seek and record complaints about foodborne diseases. For example, list the telephone number of the foodborne disease (food poisoning) unit prominently under the appropriate health agency in the phone book or on the front inside cover. To be most effective, have this number monitored 24 h a day, 7 days a week either by an answering service or a telephone emergency services system. Identify medical care facilities and practitioners, and seek their participation. Direct education activities, such as newsletters and talks at meetings, to stimulate participation in the program. Encourage food and tourist industries to report complaints of suspected foodborne illness. Some jurisdictions are exploring the use of social media to inform the public to communicate relevant information regarding public health, including food recalls and foodborne disease outbreaks. Consider developing such a strategy in your agency. Also, encourage private and hospital laboratories to report isolations of bacteria (e.g., *Escherichia coli* O157:H7, *Salmonella, Shigella, Vibrio cholerae*), viruses (e.g., norovirus and hepatitis A virus), parasitic agents (e.g., *Trichinella spiralis, Cyclospora cayetanensis, Giardia lamblia*), and other agents that may be foodborne. Develop regulations to require clinical laboratories to submit cultures of *Salmonella, Shigella, E. coli* O157:H7, and *Listeria* to state public health laboratories for further characterization. Develop a protocol for notification and coordination with agencies that might cooperate in investigation activities, including 24-h, 7-day-a-week contacts. Notify and coordinate with state/provincial, district, and national agencies that have surveillance and food regulatory responsibilities, and other national and international health agencies, as appropriate.

Collaborate with the public health laboratories and receive pulsed-field gel electrophoresis (PFGE) data on a regular basis. Such information may be useful in identifying clusters and potential foodborne outbreaks.

Assign Responsibility

Delegate responsibility to a professionally trained person who is familiar with epidemiologic methods and food safety to direct the surveillance program, report to appropriate agencies and public health partners as needed, take charge when foodborne and enteric disease outbreaks are suspected, and handle publicity during outbreaks. Delegate responsibility to others who will carry out specific epidemiologic, laboratory, and on-site investigations. If an intentional contamination of food event is suspected, law enforcement agencies may become the lead agency responsible for the investigation. Ensure that this person has methods of communicating regularly with other local state and national foodborne investigators and receiving notices about foodborne illnesses and outbreaks, e.g., national foodborne listservs.

Establish an Investigating Team

Establish a team of epidemiologists, sanitarians, public health inspectors, microbiologists, nurses, physicians, public information specialists, and others (e.g., toxicologists) as needed. Free flow of information and coordination are essential among those participating in foodborne and enteric disease surveillance and investigation, particularly when several different agencies are involved. Food-related complaints are usually directed to health departments or food regulatory agencies, but perhaps to different jurisdictions. When a food is suspected of being intentionally contaminated leading to illnesses, contact emergency response and/or law enforcement officials for their involvement. Therefore, it is essential that these complaints be recorded by the agency receiving the alerts. Whenever possible, share the information with participating parties or nearby jurisdictions by rapid means such as electronic mail or fax. Give top priority to illness and outbreak investigations.

Train Staff

Select staff members to participate in the foodborne disease surveillance program based on their interest, education, and ability. Inform them of the objectives and protocol of the program. Emphasize the value of disease surveillance in a food safety program. Assign them to investigations and encourage the use of epidemiologic information and approaches in their routine disease surveillance and prevention activities. Develop their skills so that they can carry out their role effectively during an investigation and teach them procedures to interpret data collected during investigations. Conduct seminars routinely, and during or after investigations, to update staff and keep agency personnel informed. Train office workers who receive calls concerning foodborne illnesses to give appropriate instructions. Those who participate in the investigation will learn from the experience and often are in a position to implement improvements after the investigation is completed.

Assemble Materials

Assemble and have readily available kits with forms and equipment as specified in Table A. Restock and maintain kits on a schedule recommended by laboratory staff to ensure their stability and sterility. Verify expiration dates, and use kits before this date or discard. Assemble a reference library, including e-mail and internet addresses, on foodborne and enteric illnesses and control measures; make it available to the staff. Update this library with information on foodborne pathogens and investigation techniques from articles in scientific journals, new books, manuals, and internet searches. (See Further Readings for suggestions.)

Investigate Outbreaks

Investigating an outbreak involves:

- Receiving notification of illnesses that might be foodborne
- Interviewing ill persons and persons at risk who remained well
- Making epidemiologic associations from initial information
- Forming hypotheses

  Further investigation to confirm or refute the hypotheses includes:

- Collecting clinical specimens and food samples
- Conducting an on-site investigation to determine the source and mode of contamination, survival, and/or proliferation of the etiologic agent
- Characterize further the etiologic agent by various typing schemes, e.g., serology
- Perform DNA profiling of isolates from clinical specimens and food samples to determine the extent of the outbreak, e.g., by PFGE.

Act on Notification of Illness

Prompt handling and referral of food-related complaints, rapid recognition of the problem, and prevention of further illnesses are the foundations of a successful investigation. This first contact with the public is a vital aspect of an investigation. As indicated earlier, any action respecting a potential deliberate contamination of food will generate a specific approach to further action.

Receive Complaints or Alerts

Upon receiving a complaint or an alert about a food or illness potentially attributed to food, record the information on Form A. An alert or complaint relating to food may involve foodborne illness, food spoilage, adulteration of a product, mislabeling, or an unsanitary establishment. Alerts also can be initiated by reports from physicians, by reports of foodborne pathogens isolated by laboratories, by calls to poison control centers, and by reports of treatment given in either hospital or private emergency rooms, or by emergency squads. Alerts may also include an increase in a particular PFGE pattern from clinical isolates. An investigation may be initiated to determine if there is a common food exposure among patients with the PFGE pattern. The pattern may be compared with similar PFGE patterns in the PulseNet databases to determine if there are similar occurrences of the pattern in food and clinical isolates nationwide or internationally. The form provides information upon which to decide whether an incident should be investigated. It is not difficult to fill out and can be completed by a public health professional or trained office worker.

Assign a sequential number to each complaint. If additional space is needed to record information, use the reverse side or attach additional sheets. Always ask the complainant to provide names of other persons at the event under suspicion, whether or not ill, and names of any other persons who are known to be ill with the same syndrome. Follow up by contacting these additional persons.

It is of paramount importance to collect clinical specimens and a sample of the suspect food or water as soon as possible. See section *Obtain Clinical Specimens*, (page 15) and *Collect Samples of Suspect Foods* (page 42) for detailed procedures. Tell the complainant to collect stool or vomitus specimens from ill persons using a clean utensil in a clean jar or plastic bag and to seal tightly and label clearly with name of ill person and date. Samples should be collected in containers and in accordance with the specific laboratory procedures where they will be submitted. Then, put all specimen containers into either a paper or plastic bag, label, and store in a refrigerator (but do not freeze). Emphasize the need to retain a sample of all suspect foods in their original container or package, if practical. Otherwise, instruct this person to put a half pint (250 g/mL), if that much is available, or otherwise all of the remaining food, into a clean container that has been boiled or into an unused plastic bag. Also, put this jar or bag into either a paper or plastic bag, label with name of facility, food, and date, and refrigerate but do not freeze. Instruct the ill person to hold all clinical specimens and food samples until the health agency evaluates the epidemiological evidence and arranges, if necessary, to collect them. If it is determined that the specimen or sample is not necessary, notify the complainant and advise on proper disposal of the material. Collect clinical specimens and food samples as soon as practicable and according to the recommendations given in this text.

If there is a cluster of cases, monitor reports from physicians, complaints about food, or records of laboratory isolation of enteric pathogens that may suggest outbreaks of disease or contributory situations.

Log Alert and Complaint Data

Extract key information (see * and † entries) from Form A and enter it onto Form B.

Record time of onset of the first symptom or sign of illness, number of persons who became ill, predominant symptoms and signs, and name of the food alleged to have caused the illness, if stated. Also, enter names of places or common gatherings at which the stricken person ate during the 72 h before onset of illness and other pertinent information. Under "history of exposures" column, use appropriate abbreviations to indicate the reported information. Under "comment," enter notations of type of agent isolated, results of specimen tests, places where food or water was consumed during travel, names and locations of restaurants or other foodservice facilities, and other pertinent information including hospitalization, occupation, or place of employment. At this phase of the investigation, it probably will not be known whether the illness is foodborne, waterborne, or spread person-to-person. This log can be kept either in hardcopy or in electronic format. See Table 1 (pages 9–10) as an example of a log.

Interpretation of Table 1:

- Entry 101 – Get further details on the patient's symptoms and seek other cases. The report of foreign travel suggests an infection that may have been acquired outside the country. Follow-up of such cases may identify an outbreak of international scope. If so, inform state/provincial and national authorities concerned with surveillance of foodborne disease about the situation.
- Entry 102 – Possibly foodborne but could be food from either Speedy Foods or Nitestar Club or elsewhere; look for and log possible cases with time/place associations.
- Entry 103 – Initiate investigation; 12 cases suggest an outbreak that has a common time/place association.
- Entry 104 – Refer complaint to agency or department concerned with food quality.
- Entry 105 – Vomiting suggests an illness of a short incubation period. There may be an association with the meal eaten prior to vomiting.
- Entry 106 – The syndrome is similar to that of Entry 105. The names Joe's and Jo's are similar; clarify.
- Entry 107 – Look for other cases having the same infection.
- Entry 108 – Look for other cases with similar signs and symptoms with the same place/event association.
- Entry 109 – A second isolation of *Salmonella* within a few days, but a different serotype. Association with Entry 107 is unlikely.
- Entry 110 – The illness is similar to that caused by norovirus which can be transmitted by oysters. Look for other cases with similar signs or with a place association.
- Entry 111 – The second complaint associated with eating at Joe's Diner is a possible common place association (see Entry 105); investigate.
- Entry 112 – Report situation to agency or department responsible for regulating food processing and monitoring adulterated food.
- Entry 113 – This syndrome also suggests norovirus illness. There may be a possible association with the place where the oysters were harvested and there may be an association with Entry 110; investigate.
- Entry 114 – This illness may be associated with Entry 107. Both persons are ill with salmonellosis caused by the same serotype of *Salmonella, S.* Chester; investigate.
- Entry 115 – The signs and systems suggest botulism; investigate immediately and collect clinical specimens and food samples.

Review the log each time an entry is made and also each week to identify clusters of cases and/or involvement of a common food or place of eating that might otherwise go undetected. If your agency has district offices or if there are nearby jurisdictions (as in metropolitan areas), periodically send copies of log sheets to a

Table 1 Foodborne, waterborne, enteric illness and complaint log.

| Complaint | | Illnesses | | | | Food | | Water | | | | |
No.	Date	Type[a]	Onset date	No. ill	Predominant symptoms/signs	Alleged/ Suspected	Where eaten within 72 h	Where ingested within 2 week	Where contacted within 2 week	Source[b]	History of exposure[c]	Comments
101	8/16	I	8/15	1	Diarrhea			Redguard		C	IT	*G. lamblia* isolated from stool specimen
102	8/20	I/UE	8/18	1	Diarrhea	Hamburger	Speedy Foods	Dixon		C		
103	8/21	I	8/21	12	Diarrhea	Roast beef	Nitestar Club	Jonesville		C		
							Dixon Child Care Center	Dixon		C	CC	
104	8/23	CF	8/23	0		Corn		Daulton				Swollen can, Brand W, code LM 308
105	8/24	I	8/23	1	Vomiting	Ham	Joe's Diner	Dixon				
106	8/23	I	8/23	1	Nausea, vomiting	Cold cuts from deli	Jo's Market	Dixon		C		*Salmonella* Chester
107	8/26	I	8/24	1	Diarrhea, fever, vomiting			Plainville		W		*S.* Chester isolated from stool
108	8/26	I	8/25	1	Diarrhea, fever		Church supper	Dixon		C		*Salmonella* Anatum

(continued)

Table 1 (continued)

109	8/27	I		8/25	1	Diarrhea, chills			Midvale		*S. anatum* isolated from stool
110	8/27	I		8/26	1	Vomiting, diarrhea	Oysters	Fred's Cafe	Dixon	C	
111	8/29	I		8/28	1	Vomiting		Joe's Diner	Clear Falls	W	
112	8/29		CF	8/28	0		Baby cereal		Dixon	C	Metal fragments found in Brand X, Lot JK201E
113	8/30	I		8/28	1	Nausea, vomiting, headache	Oysters	Ralph's Seafood	Dixon	C	Canned by Food Processors Inc. Lot L36105F
114	9/1	I		8/29	1	Diarrhea, fever, vomiting		Dixon Arts Festival	Dixon	C	*S. Chester* isolated from stool
115	9/2	I		8/29	1	Blurred and double vision, paralysis			Dixon	C	Hospital reported assisted breathing

[a]Type complaint: I illness; CF contaminated/adulterated/spoiled food; UE unsanitary food establishment; DW poor quality drinking water; RW poor quality recreation water; MP complaint related to media publicity; D disasters; O other

[b]Water source: C community; NC noncommunity; W well; B bottled; S/L stream/lake

[c]Exposure history: DT domestic travel (out of town, within country); IT international travel; CC child care; CI contact with ill person outside household or visitor to household; An Exposure to ill animal. C contact with ill person in household

central coordinating office (e.g., weekly or when there are 10–20 entries). Reports of current illness levels should include historical information on illness trends in the community so that new data can be considered in the appropriate context. Report to your supervisor if you suspect any time, place, or person associations and take steps to initiate an investigation.

Refer Complaint to Proper Agency

Refer complaints that fall outside your agency's range of operations to the appropriate authority, such as the Department of Health, Department of Agriculture, etc., and indicate the action taken in the disposition box on Form A (page 139). Develop a working relationship with such authorities so they will reciprocate in situations that may be associated with illness. Often an investigation requires efforts of more than one agency. Cooperation and prompt exchange of information between agencies are vital.

Prepare for the Investigation

Assign responsibility for an investigation to the person heading the investigative team if this was not done when the surveillance protocol was established. Delegate sufficient authority and provide resources to this person so that the task of investigation can be accomplished. Inform everyone working on the investigation that findings are to be reported to this person.

Before beginning the investigation, check the supply of forms and the availability of equipment suggested in Table A (page 108). Obtain any needed materials or additional equipment.

If the alert or complaint suggests a possible outbreak, inform laboratory personnel of the type of outbreak and estimated quantity and arrival time of clinical specimens and food samples to be collected. This information will give laboratory managers time to prepare laboratory culture media, prepare reagents, and allocate personnel. Consult laboratory personnel about proper methods for collecting, preserving, and shipping food samples and clinical specimens if such information is needed. Get appropriate specimen containers from them.

Verify Diagnosis

An ill person or family member, physician, or hospital staff member may report suspected cases of foodborne illness. Whatever the source of the report, verify the diagnosis by taking a thorough case history and, if possible, by reviewing clinical information and laboratory findings. (This analysis can be further substantiated by detecting suspected etiologic agents in foods.) This verification is done in consultation with medical professionals.

Get Case Histories

When a complaint involves illness, complete Form C1–2 (pages 141–142) either at the time of initial notification or during a personal visit or a telephone call to the person reported to be ill. Use this detailed interview approach with every person who has been identified in the initial complaint or alert, even though some may not have been ill. Be aware that potential cultural and language barriers can make interviews difficult. A different interviewer may be needed to accommodate these barriers. Continue this until sufficient information is obtained to decide whether there is, indeed, an outbreak of foodborne illness. From persons who are at risk of illness but who remained well, also get 72-h food histories and information about their activities in common with the ill persons. Information from these persons is as important to make epidemiologic associations as it is from the cases.

When it is apparent that an outbreak has occurred (for definition of an outbreak) and a specific event has come under suspicion, substitute Form D1–2 (pages 143–144) for Form C. Form D can be used initially in many routine foodborne illness outbreak investigations where it is obvious that a common-source outbreak has occurred or when all of the ill ate food together (e.g., school lunches, church supper, or banquet); at the same place (e.g., restaurants); or event (e.g., festival). This will simplify recording, because most affected persons will give similar information. At this time, notify the district, state, or provincial epidemiologist about the outbreak.

If a specific pathogen (e.g., *Salmonella, E. coli* O157:H7, hepatitis A virus) has been identified as the etiologic agent, consider developing a form for recording relevant information. Many state/provincial or national public health agencies have standard forms tailored to specific pathogens. Include signs and symptoms of the illness and other clinical information, the etiology of the agent, and usual methods of transmission. Computer programs (e.g., Epi Info™) can aid in the design of such forms.

Upon contact with the affected person, identify yourself and your agency and explain the purpose of the visit or call. A professional attitude, appropriate attire, friendly manner, and confidence in discussing epidemiology and control of foodborne illnesses are essential for developing rapport with affected persons or their families and in projecting a good image of the investigating agency. Keep in mind that you are not interviewing someone you inspect or regulate, but that you are providing a service to the affected person. Exhibit genuine concern for persons affected and be sincere when requesting personal and confidential information. Communicate a sense of the urgency of the investigation, and emphasize that their participation will make a positive contribution for the control and prevention of foodborne illness. Parental consent must be obtained before interviewing children under 18 years of age. In some locations, consent from the affected person's physician may also be required.

After asking open-ended questions about the person's food exposures and illness history, follow up with more specific questions to fill in the details and better ensure a thorough recall. Base your level of communication on a general impression of the person being interviewed, considering information about age, occupation, education, or socioeconomic status. Tact is essential! Use either Form C or Form D, as appropriate, as a guide. State questions so that the persons being interviewed will describe their illnesses and associated events in their own words. Try not to suggest answers by the way you phrase questions.

Fill in Form C1–2 (if appropriate) and take additional notes during the interview. Ask specific questions to clarify the patient's comments. Think questions through before conducting the interview. Realize that people are sometimes sensitive to questions about age, sex, special dietary habits, ethnic group, excreta disposal, and housing conditions. Nevertheless, any or all information of this type can be relevant. Word questions thoughtfully when discussing these characteristics and habits. Such information can often be deduced from observations. If doubt remains, confirm your guesses by asking indirect questions. Information on recent travel, gatherings, or visitors may provide a clue to common sources or events that would otherwise be difficult to pinpoint. Review known allergies, recent immunizations, recent changes in the patient's medical status, and similar information. Remember that the agents for foodborne disease can also be spread by other means such as consuming contaminated water, person-to-person, in a child care center, animal-to-person in petting zoos, or spread through the walk-in-spray fountain or kiddie pool.

As persons describe their illnesses, check boxes next to appropriate symptoms or signs on Form C1. Do not ask about all symptoms or signs listed; however, ask about those marked with an asterisk if the ill person does not mention them. If there are questions, explain symptoms to the patient in understandable terms. The symptoms and signs in the first two columns of Form C1 are usually associated with poisoning or intoxication, although some occur during infections. Those in the third, fourth, and fifth columns are usually associated with enteric infections, generalized infections, and localized infections, respectively. Those in the last column are usually associated with disturbance of the central nervous system. Diseases in any category will sometimes be characterized by a few symptoms and signs listed in the other columns, and not all signs and symptoms occur for any one ailment or for all persons reporting illness. If an illness seems to fall into one of these categories, mention other symptoms in the category and record the patient's response.

Whenever possible, use physician and hospital records to verify signs and symptoms reported by patients. Clinical data may strengthen or dismiss the possibility of foodborne illness. Before contacting a physician or a hospital, become familiar with laws and codes relating to medical records to ensure that you have legal access to these records. Legal release forms may be necessary to obtain some records. Do not distribute names of patients, their other personal identities (e.g., address, phone number), or their clinical information to unauthorized persons.

Signs and symptoms will sometimes give a clue to the transmission route by indicating the organ systems affected. (See Table B, page 109–126, for listings of typical signs and symptoms and incubation periods of specific diseases.) If the early and predominant symptoms are nausea and vomiting, ask about foods or beverages ingested within the last 6 h. In these situations, suspect a bacterial (e.g., staphylococcal intoxication, *Bacillus cereus* gastroenteritis) or chemical poisoning. High-acid beverages (e.g., carbonated beverages, fruit drinks) tend to leach metallic ions from pipes and containers. If diarrhea and abdominal cramps predominate without fever, be suspicious of foods eaten 6–20 h before onset of illness. Primary concerns are *Clostridium perfringens* and diarrhetic *B. cereus*. If diarrhea, chills, and fever predominate, be suspicious of foods and beverages ingested 12–72 h before onset of illness for salmonellosis, shigellosis, *E. coli* infections, and norovirus gastroenteritis.

Gather information about all foods eaten at least 72 h, and water ingested 2 weeks, before onset of illness. The food, even the meal, which precipitated the illness might not be obvious; persons often have difficulty in remembering all foods and components (i.e., sauces, spices) eaten and ice or water ingested during these intervals. Therefore, if the person does not remember specific foods eaten, ask about usual foods eaten and events attended during this interval. This may stimulate recall. The entries begin with the day of illness, followed by the previous 2 days. If the illness, however, began early in the day or before any of the listed meals, modify the entries on the form so that the 72-h history can be completed in the space provided on the form. If the incubation period is 3 days to a week in duration, use additional copies of Form C2 and modify day or day before subtitles.

If the incubation period averages 1–2 weeks, consider typhoid fever, cryptosporidiosis or giardiasis. Diseases with incubation periods exceeding 2 weeks (e.g., hepatitis A and E) can be handled as special cases for which longer histories would be sought. In these situations, omit the 72-h food history and enter other potential vehicles (e.g., water and shellfish if hepatitis A is suspected). Enter this information in the row that cites the history of ingesting suspected food.

Other microorganisms not listed in Table B are potentially spread by food, but they are rare or not yet identified as being foodborne. These may be opportunistic pathogens, particularly for highly susceptible and immunosuppressed persons. Further investigation is needed to confirm their role in the spread of foodborne diseases.

Get information about the usual places that the ill persons eat, sources of water and ice, and places where food was consumed during the past 72 h. Ask about other risk factors for enteric illness, such as contact with young children and child care centers, animal contact, ingestion of raw foods of animal origin, and usual food preference habits. For persons who have been traveling, ask them where (both cities and rural areas) they have traveled during the incubation period of suspected agents. Consider both domestic and international travel. Determine if they ate food at unusual events or consumed foods different from their usual diet (e.g., at a festival or while on vacation). This information sometimes provides clues to common sources or to events that otherwise would be difficult to discover. Record the information on Form C.

In a protracted outbreak or when investigating an outbreak of a disease with a long incubation period, expect recall to be poor. In this situation, obtain from ill persons and others at risk a listing of their food and beverage preferences, amounts usually ingested, or their purchases of these items within the range of the incubation period of the suspected disease. As a guide, draw up a list of either foods and beverages that are commonly consumed by the affected group or foods and beverages previously identified as vehicles of the suspected disease under investigation.

Summarize data from all copies of Form C1–2 on Form D1–2. Form D allows rapid review of all exposed persons (ill or not ill) and serves as a basis for analyzing the data.

Obtain Clinical Specimens

Diagnosis of most diseases can be confirmed only if etiologic agents are isolated and identified from specimens obtained from ill persons. Get specimens from the ill persons to confirm an etiologic agent.

- In large outbreaks, obtain fecal specimens from at least ten persons who manifest illness typical of the outbreak.
- In smaller outbreaks, obtain specimens from as many of those ill and those at risk as practicable, but from at least two, and preferably ten, ill persons.
- Try to collect specimens before the patient takes any medication. If medication has already been taken, collect specimens anyway, and find out the kinds and amounts of medicine taken and the time that each dose was taken.
- Also get control specimens from persons with similar exposure histories who did not become ill.

Obtain clinical specimens at the time of the initial interview during acute illness or as soon as practicable thereafter. Apart from the fact that people are more likely to cooperate while they are ill, some pathogens or poisonous substances remain in the intestinal tract for only a day or so after onset of illness (e.g., *C. perfringens, B. cereus*). If the patient is reluctant to provide a fecal specimen explain that the specimen will be tested to identify the causative agent and compare it to any agent recovered from the food.

If a disease has already been diagnosed, collect specimens as listed in Table B. If a disease has not yet been diagnosed, choose specimens that are appropriate to the clinical features. Laboratory information obtained from the first patients may be useful to physicians in treating cases detected later.

Some pathogens (e.g., *Salmonella,* parasites) may be recovered for weeks after symptoms have abated. If applicable for the disease under investigation, take specimens even after recovery because some etiologic agents may remain in low numbers, and changes in serologic titers can be detected.

Before collecting specimens, review Table C and, if necessary, get additional instructions from laboratory personnel and seek their advice on how to preserve the stool specimens if you cannot deliver them to the laboratory immediately. Many public health agencies have special fecal specimen kits. Demonstrate to the patient how to use the materials in the kit, how to complete the form in the kit and how to mail it if you are not going to pick it up. If mailing specimens, make sure that you are aware of the regulatory requirements that may apply to the transport of infectious material.

Stool specimen containers for intestinal parasite examination are not suitable for bacterial or viral examinations because they ordinarily contain a preservative, such as formalin or polyvinyl alcohol. If an inappropriate transport medium is used, a specimen can be rendered unsuitable for laboratory examination.

Feces. If the patient has diarrhea or is suspected of having had an enteric disease, obtain a stool specimen (preferred specimen) or a rectal swab. Instruct patients to provide you with their own specimens by one of the following means.

1. If practicable, give the patient a stool specimen container or with a wooden or plastic spoon or a tongue depressor. A clean container available in the home (e.g., a jar, or disposable container that can be sealed) and a clean plastic spoon or similar utensil can be used if laboratory containers are not available.
2. Label the specimen container with the patient's name age/date of birth and date of collection.
3. Collect the stool specimen by one of the following methods:
   a. Put sheets of plastic wrap or aluminum foil under the toilet seat and push them down slightly in the center, but not so far as to touch the water in the bowl. Sheets of paper can be tacked on the rise of a latrine and pushed down to form a depression in which to catch feces. Take care to ensure that toilet cleaning chemicals and other microorganisms in the toilet bowl do not contaminate the fecal specimen. After defecating, use a clean spoon or other utensil to transfer about 10 g of feces into a specimen container or other clean container.
   b. Defecate directly into a large clean dry container or bedpan. Use a clean spoon or other utensil to transfer about 10 g or the size of a walnut of feces into a specimen container or other clean container.
   c. Scrape feces off a diaper with a clean spoon or other utensil to transfer about 10 g of feces into a specimen container or other clean container.
4. Collect fecal swabs by twisting the cotton-wrapped end of the swab into the stool obtained in one of the ways described above. Follow instructions given in Table C. If necessary, use fecal-soiled toilet paper or cloth diaper and twist a swab into the top of feces. Take care to ensure that there is no carryover of toilet paper as they are impregnated with barium salts which are inhibitory to some fecal pathogens.

Dispose of excess fecal material into the toilet and carefully wrap all soiled articles (e.g., by placing them inside two plastic bags) and dispose of in domestic waste. Check that the specimen container is tightly sealed and properly labeled and place into a clean outer plastic bag (special zip lock bags for clinical specimens, if available). Store the specimen in a cool place, preferably at 4°C to await pick-up or despatch. DO NOT FREEZE.

Feces from Rectal Swabs. Collect rectal swabs by carefully inserting the swab approximately 1 in. (2.5 cm) beyond the anal sphincter. Gently rotate the swab. Fecal matter should be evident on the swab.

Vomitus. If the person is vomiting or subsequently does so, arrange to collect vomitus. Tell the patient to vomit directly into a sterile specimen container or a plastic bag. Otherwise, transfer some vomitus from a clean receptacle into the container with a clean spoon. Refrigerate, but do not freeze, this specimen until it can be picked up or delivered to the laboratory.

Blood. Take blood if a patient has a febrile infection or when infectious agents or botulism are suspected (see Table B). Blood specimens are collected for:

* Bacterial culture
* Detection of antibodies to specific agents
* Detection of certain toxins

Before collecting specimens, get additional instructions from laboratory personnel and seek their advice.

Blood should be obtained by an appropriately trained and accredited person (check appropriate laws). Collect blood during the acute phase of illness, as soon as the febrile patient is seen (within a week after onset of illness) and, if comparing of serologic titers, again within 6 week (usually 2–4 week later) during the convalescent phase. Draw 15 mL of blood (from an adult) or 3 mL (from a child) or 1–2 mL (from an infant). If possible, collect the blood from the same patients from which stool specimens were obtained if both specimens are to be examined.

Label tubes and vials at every step of serum transfer. Do not freeze whole blood because the resultant hemolysis interferes with serologic reactions.

Blood for culture (for pathogens such as invasive *Salmonella* species, *Vibrio vulnificus, Bacillus anthracis*)	Inoculate freshly collected blood into culture bottle supplied by the laboratory
Blood for detection of – antibodies (to pathogens such as *Salmonella* Typhi, hepatitis A virus, *Toxoplasma gondii*) – toxins (such as botulinal toxin)	Collect into a sterile syringe or evacuated sterile tube that does not contain anticoagulants. If practicable, centrifuge the blood at 1,000 rpm for 10 min; pour off the serum into small screw-cap vials and store at approximately −18°C (0°F). If the serum cannot be separated immediately, rim the clot with a sterile applicator stick and refrigerate at approximately 4°C (40°F) to get maximum clot retraction if the specimen is to be stored unfrozen overnight. If centrifugation cannot be done, store the blood specimens in a refrigerator until a clot has formed, then remove the serum and transfer it with a Pasteur pipette into an empty sterile tube. Send only the serum for analysis

Urine. Instruct patients to collect urine in the following manner. Clean the area immediately around the urethral orifice with a paper pad that has been premoistened with 4% tincture of iodine or other appropriate antiseptic. Then begin to urinate into a toilet and collect 30 mL (about 1 oz.) of midstream urine into a sterile bottle. Use either a second antiseptic-moistened pad or an alcohol-moistened cotton ball or tissue to clean any drops from the top or side of the bottle.

Other Instructions. Follow applicable instructions given in Table C. Before or immediately after collecting clinical specimens, use waterproof permanent markers to label each container with the patients name, complaint number, case identification number, specimen number, date and time of collection, tests requested, and other appropriate information. Tightly seal all containers. Complete Form E, *Clinical Specimen Collection Report* for each specimen. The complaint number, case identification (ID) number, and specimen number must be entered on each report so that laboratory results can later be correlated with other data. Record on either Form C or D the type of specimen collected, and submit both the specimen and a copy of Form E (page 145) to the laboratory. Send a copy of the laboratory report to the patient's physician or call if urgent.

Pick Up Food Samples and Containers that the Patient Collected

If the patient/case or other household member collected any suspected foods as instructed during initial contact, label containers with the complaint/outbreak and sample numbers. Also, collect leftovers of foods that were eaten in the last 72 h if home-prepared foods are under suspicion. Proceed as instructed in *Collect Samples of Suspect Foods* section (page 42), and Form F, *Food Sample Collection Report* (page 146). Record conditions of collection as called for on the forms. If the clinical data or the food histories associate the illness with a food, caution persons not to use the remaining stocks of suspect foods until the investigation is complete. If testing a large number of samples is impractical, examine the most highly suspect food or foods first. Hold the other samples refrigerated in the laboratory for testing later, if necessary.

If the suspect food is a commercial product, obtain the original container (e.g., can, label, and lid), if feasible, even if it has been discarded. An empty container can be used to identify the processor and micro-leaks, or rinsings from these containers can be used to detect microorganisms or toxins. Check the original package or container for a code number that can be used to identify the place and time of processing. Record on Form F pertinent data such as the time and place of purchase.

Develop a Working Case Definition

Develop a working case definition to classify exposed persons as cases or non-cases. Start with the most pronounced signs and/or symptoms (such as diarrhea and vomiting) rather than nonspecific symptoms (such as nausea, abdominal cramps, or malaise). For example, in an outbreak of gastroenteritis, a case might be defined as a person who was at risk and developed diarrhea within a specified period of time. Diarrhea must be defined, perhaps as three or more loose or watery stools within 24 h. An investigation might also be initiated as a result of clinical specimens yielding the same pathogen. A case definition, which is developed later in the investigation, might include either a person having signs and/or symptoms characteristic of a likely syndrome within a period of time or a person from whom a specific pathogen was isolated (See Table B).

Sometimes the first symptom or sign provides a clue to developing a case definition. Use Tables B and D as guides for making case definitions. Compare each newly identified case with the definition to see whether it is part of the outbreak. Cases may be classified into one of three categories:

- *A confirmed case* is a person who has signs and symptoms that are clinically compatible with the disease under consideration, and for which there is either (a) isolation of an etiologic agent from (or otherwise identified in) an appropriate specimen from the patient, or (b) serologic evidence of a fourfold or greater rise in convalescent antibody titer. A confirmed case must also have possible exposure to the etiologic agent within the incubation period of the disease.
- *A presumptive case* is a person who has signs and symptoms that are clinically compatible with the disease under consideration, and for which there is laboratory evidence of infection (e.g., an elevated antibody titer, but less than a fourfold increase), but the etiologic agent has not been found in specimens from patients or no specimens were collected. A presumptive case must also have possible exposure to the etiologic agent within the incubation period of the disease.
- *A suspected case* is a person who has signs and symptoms and incubation period that are clinically compatible with the disease under consideration and history of possible exposure, but laboratory evidence is absent, inconclusive, or incomplete.

It is not essential, however, to classify cases into these categories. Do so only if it aids in developing a final case definition or in making comparative analyses of data.

A secondary case is a person who became infected from contact with a primary (outbreak-associated) case or from a vehicle contaminated by a primary case. Onset of illness for secondary cases typically is one or more incubation periods after the outbreak-associated cases.

Consider doing analyses using case definitions of both confirmed and combined confirmed, presumptive, and highly suspect cases, and compare the results. The ultimate case definition has a tremendous impact on ability to make illness and exposure associations and to calculate probability of these associations.

Make Epidemiologic Associations

Make a preliminary evaluation of the data collected as soon as possible. If you decide that there is an outbreak, use the information that you have collected to develop a hypothesis about the causal factors.

Make Time, Place, and/or Person Associations

A time association exists if the times of onset of similar illnesses are within a few hours or days of each other. Place associations exist when persons (a) purchase foods from the same place, (b) eat at the same establishment, (c) attend the same event, or (d) reside in a place common to all. Person associations suggest a shared personal characteristic, such as being of the same age grouping, sex, ethnic group, occupation, social group, or religion. Once some of these associations become obvious, question other persons who could be at risk because of their time, place, or person association with the ill persons.

Decide Whether an Outbreak Has Occurred

An outbreak is an incident in which two or more persons have the same disease, have similar clinical features, or have the same pathogen with the same PFGE pattern – thus meeting the case definition – and there is a time, place, or person association among these persons. A foodborne outbreak is one that is traceable to ingestion of a contaminated food. A single case of suspected botulism, mushroom poisoning, ciguatera or paralytic shellfish poisoning or other rare disease (e.g., *Vibrio vulnificus*), or a case of a disease that can be definitely related to ingestion of a food can be considered an incident of foodborne illness and warrants further investigation.

Sometimes an outbreak of foodborne disease can be determined from an initial report simply because of the number of persons displaying certain signs and symptoms at about the same time. Many complaints, however, involve illness in only one or a few persons. It is often difficult to decide whether ingestion of a particular food and the onset of illness were associated or coincidental. Certain diseases that are highly communicable (e.g., shigellosis and epidemic viral gastroenteritis) may result in secondary infections, person-to-person spread or subsequently contaminated food or water. Other common source outbreaks, such as those caused by water or cross contamination from diapers in child care centers, may also simulate foodborne outbreaks. If complaints are received from several persons who have eaten the same food or who have eaten at the same place, food is likely to be involved. Routinely reviewing the *Foodborne, waterborne, enteric illness and complaint log* (Form B, page 140) for similar complaints can often be useful in detecting time, place, or person associations. If the agent involved is *Salmonella, Listeria,* or *E. coli* O157:H7, the isolate should be submitted to the state/provincial or national public health laboratory for inclusion into national databases that contain specific phage types or serotypes or genetic markers such as pulsed-field gel electrophoresis (PFGE) patterns (i.e., PulseNet). The pathogen in your outbreak can be compared with these databases. If the organism has been found recently in humans or foods, you may have a link to additional cases or the food that is causing this outbreak. An investigation also may proceed from an alert of an intentional contamination of a food item. This is further discussed on page 99.

Formulate Hypotheses

From time, place, or person associations that have been established or suggested by the investigation including patient food and nonfood exposures, formulate hypotheses to explain (a) the most likely type of illness, (b) the most likely vehicle involved, (c) where and the manner by which the vehicle might have become contaminated, and (d) other possible causal relationships. (The section, *Analyze Data*, page 50 describes calculations that can aid in the formation of these hypotheses.) If the hypothesis relates to foodborne illness, follow the instructions given in this manual. If the hypothesis relates to waterborne illness, follow the instructions given in *Procedures to Investigate Waterborne Illness*. If the hypothesis relates to other sources, seek other appropriate guidelines.

Expand the Investigation

Test hypotheses by obtaining additional information to confirm or refute their validity. Do this by case–control or cohort studies, additional laboratory investigations, and on-site investigations (e.g., food preparation review).

Obtain Assistance

If an outbreak investigation requires resources beyond your agency's capacity, request assistance from other health professionals. It is desirable to have a team including, if feasible, an epidemiologist, a microbiologist, a physician, a sanitarian, a toxicologist, and others qualified to undertake a detailed foodborne illness investigation. Such personnel can usually be provided by local, state/provincial, or national agencies concerned with health, food, environment, fisheries, or agriculture, depending on the expertise needed. For events suspected to arise from intentionally contaminated food, contact emergency response or law enforcement agencies (page 101).

Find and Interview Additional Cases

Continue to search for and interview ill persons who have had time, place, or person associations with the identified cases. (See the section *Make Time, Place, and/or Person Associations*, page 19)

Review recent complaints in the complaint log for possible outbreak-associated cases (Form B). Contact other nearby health agencies, hospital emergency rooms, and local physicians to discover other epidemiologically related cases. Call previously contacted persons to see whether they know of anyone else who has become ill or who has a common association suggested by data in the log. Obtain reservation lists, computer records, and credit card slips from foodservice establishments to attempt to find others who were exposed.

The illness you are investigating may be part of a larger multijurisdictional outbreak, and therefore communicate with adjoining local and state agencies to learn if they are seeing similar illnesses. State or provincial public health agencies can check reportable disease records and state/provincial public health laboratories can start looking for clusters in isolates that they are characterizing. For outbreaks where intentional contamination of food is suspected or confirmed, public health and law enforcement agency officials may conduct the investigation jointly (page 101).

If it becomes apparent that an outbreak is associated with a specific event or one meal, use Form D for recording information. At this stage of the investigation, interviews can be expedited by reviewing the event or meal itself to stimulate each person's recall. Ask about specific symptoms and signs that are common to the syndrome, the time of eating the suspect meal or food, and the time of onset of illness. Mention each food served at the event or meal to which the person may have been exposed, and ask each person (whether a case or a well person at risk) which of the foods they ate. Also ask about foods eaten that are not listed on the questionnaire.

The number of persons to be interviewed depends on the number exposed and the proportion of them who are probably affected; if fewer than 100 persons were at risk, try to interview all of them; if several hundred are involved, interview a representative sample. Be sure to obtain clinical specimens from these cases and well persons at risk of exposure (controls). (See section *Obtain Clinical Specimens*, page 15). There may be situations where questionnaires are sent to cases and persons at risk for them to answer and return. Use either Form C or D or modified versions for this purpose. After questionnaires have been completed, summarize the data on Form D. Also, identify and interview secondary cases if they become apparent.

Find and Interview Controls

Statistical analyses of outbreak data cannot be performed without a group of non-ill persons (controls) or persons at risk who did not become ill. Controls are persons who are not ill but have other characteristics of the cases. If feasible, select two controls per case. A control may be a family member or neighbor who remained well. Controls also may be obtained by calling billing receipts, by contacting a neighbor, or by calling random phone numbers.

A common way to search by telephone for eligible controls is to add 1 to the last digit of the primary residence telephone number of the case for the first 100 telephone calls and subtract 1 from this number for the second 100 calls, and so on. Call each number only once. Discontinue the call and dial the next telephone number if any of the following apply: there is no answer; the answer is by answering machine, pager, or facsimile machine; the person answering is uncooperative, mentally or physically impaired, does not speak English, or the standard language(s) of the community, or discontinues the interview prematurely; answered by a person under 18 years of age; or the number is a place of business. Continue until two controls are identified for each case. Controls are also excluded if they have had diarrhea (for situations of enteric illnesses) within the past month or report illness or culture-confirmed infections of the agent under investigation of a household member.

Interview these persons to gather information essential to make statistical comparisons. This includes whether the control ate the meals or foods being statistically evaluated. Record this information on the same forms (e.g., C, D) as used for recording information from the cases.

Modify Procedures, as Necessary

Because no two foodborne disease outbreaks are identical, the order of the expanded investigation may not always follow the outlined sequence of procedures. Some investigative steps can usually be done simultaneously by different investigators. Although additional procedures may be required, the principles and techniques described will suffice for most investigations. Modify or develop additional forms, if necessary, to accommodate the type and amount of information that is to be collected.

If the illness you are investigating is part of a multijurisdictional outbreak, you will need to coordinate your investigation with appropriate agencies and officials at the local, state/provincial, or national level as is dictated by the scope of the outbreak.

Seek Sources and Modes of Contamination and Ways by Which the Contaminants Survived and/or Proliferated

Conduct a food processing/preparation review at the site where the suspected food was or foods were produced, processed, packaged, prepared, transported, stored, and/or served (i.e., where they may have been mishandled and/or mistreated), as applicable to the situation being investigated. The food processing/preparation review can prove or refute hypotheses developed during the epidemiological portion of the investigation. The concerns of this phase of the investigation are foods and the operations that they have undergone. Focus on source and mode of contamination and ways by which contaminants survived and/or proliferated. This phase of the investigation frequently will be at the site of final preparation of the epidemiologically implicated vehicle. However, if contamination, survival, and/or proliferation are hypothesized to have occurred before arrival at this site, initiate a traceback investigation to determine these factors (see section *Trace Contamination and Malpractices to Their Sources,* page 48).

A food preparation/preparation review is quite different from inspections conducted during routine evaluations of foodservice, warehousing, processing, and distribution facilities. If significant matters relating to food safety or quality are observed or otherwise identified during the investigation, note them and communicate them to proper authorities. Do not confuse matters of quality and esthetics with the focuses of the investigation: contamination, survival, and proliferation of infectious and toxic agents.

Seek procedural advice from the legal team in your organization if prosecution or other legal action is likely to arise from the investigation. If a food has been intentionally contaminated, the standard modes of contamination, pathogens involved, numbers present and typical pathogen food combinations involved may be very different and will require different approaches to generate the relevant data (page 102).

Plan On-Site Investigation

Identify persons responsible for operating and managing the implicated food facility before visiting the site, if feasible. Consider the types and sources of records that should be reviewed during the investigation. For example:

- Menus
- Recipes or product formulation records for foods under suspicion, with particular concern about recent changes
- Processing records (e.g., retort records, pasteurization charts)
- Operational manuals
- Hazard analysis critical control point (HACCP) monitoring records and time–temperature logs
- Previous hazard analyses and HACCP plans for products
- Process flow diagrams
- Absenteeism records
- Cleaning records
- Changes in job assignments in the establishment
- Product testing and challenge testing results
- Records of complaints on the facility or product defects
- Recent change of supplier
- Steps taken to secure products (biosecurity)

Use tact when requesting this information. In some states/provinces or countries, some of this information is protected as a proprietary right and firms may refuse to reveal it.

Alert laboratory personnel that a field investigation will be made and get suggestions for samples and specimens that should be collected (see Tables C and D). Confer with them about special analyses, transport media, and sampling procedures; also make arrangements for rapid transport and testing of samples. Gather appropriate forms (E, F, G, H, I, J) and sample collection and specimen collection equipment, preferably preassembled in a kit (see *Assemble Materials* section, page 5). Also, assemble thermometers or thermocouples and potentiometer, pH meter, and other appropriate measuring devices that might be used during the field investigations. (See Table A for further suggestions.) Ensure other laboratories submit cultures to the state public health laboratories for PFGE in a rapid manner.

Coordinate with the person who has regulatory responsibilities for the establishment under investigation. Invite this person to join the investigation team. Past inspection records may suggest problems, but be aware that outbreaks can occur, and often have occurred, in what have been considered good establishments and that previous records will seldom reflect the current situation. Routine inspection forms frequently do not emphasize critical operations whose failure may contribute to outbreaks. Furthermore, previous inspections may have been made months before when operations, foods being prepared, and personnel were different. Additionally, inspection may have been at a time of day when crucial preparation, mishandling, or mistreatment of foods did not occur, or the suspected food was not prepared or processed at that time.

At the establishment under investigation, plan to analyze all operations to which the foods of concern were subjected using Keys A–F (pages 80–97). (See explanation of the Keys with four examples on pages 39–40). Plan to stay in the establishment long enough to evaluate all suspicious processes. A full day and evening, or even longer, may be required. Do not make a routine sanitary inspection nor use the inspection form designed for that purpose. These will divert the investigation from its objectives of determining sources and modes of contamination, likelihood of microbial survival or toxin retention during processing, and opportunities for microbial multiplication.

Meet Managers

Introduce yourself to the person in charge and state your purpose immediately upon arrival at the place where the suspect food was processed or prepared, or where the implicated meal was served. Emphasize that the purpose of the investigation is to determine events or activities that contributed to an outbreak of foodborne disease so that preventive measures can be taken, and, hence, the episode can be controlled and will not be repeated. Attempt to create a spirit of cooperation as workers' attitudes toward the investigative team are influenced by a positive, communicative, working relationship between management and the investigator. Consider the position, feeling, and concerns of the manager and workers; defensive reactions are to be expected. Many factors could have contributed to contamination or bacterial multiplication before foods came under the control of the manager. Assure the manager that these possibilities, when applicable, also will be investigated. Inform the manager of the proposed activities and benefits that may be gained from the findings for training workers. Get menus, recipes, information about the products prepared, product flow, names of persons responsible for particular operations, and other relevant records (see *Plan On-site Investigation* section, page 24).

Take a preliminary look at the entire operation to observe conditions that may have contributed to the outbreak. At this time, if appropriate, take relevant measurements of critical operations before they are modified because of your investigation, and obtain samples of foods before they are discarded.

Maintain an unbiased attitude and answer questions – other than those concerning the identities of the persons whose common experiences implicated the food establishment. Do not be distracted, however, from the objectives of the visit, which are to determine source and mode of microbial or chemical contamination of the food; likelihood that pathogens survived any process designed to kill them or reduce their populations; and opportunities for growth of pathogenic bacteria or toxigenic molds. As pertinent information is obtained, take notes that can be used in a flow diagram on Form G, *Flow Process of Implicated Food* (page 147), or for litigation that may be forthcoming, and record them on Form H, *Food Processing/ Preparation History and Hazard Analysis Report* (page 148). Photographs may be a useful aid for recall and as supporting material in litigation.

Draw a Flow Diagram of Operations

Draw a separate flow chart for each food being investigated, showing each operation on a copy of Form G, *Flow Process of Implicated Food*. Use information obtained from the manager and from recipes and preparation guidelines to start this drawing. Modify it with subsequent information obtained from food workers. This is necessary because workers do not always follow prescribed procedures, and managers are not always aware of all the activities of workers. Accompany workers on a walk through the processing steps for the suspect foods from receiving and storage to shipment or serving. Identify opportunities for contamination, survival, or proliferation at each step. Compare observations made during the walk with descriptions of the process that workers and the manager gave you. Clarify and resolve any inconsistencies. In the flow diagram, each operation is represented by a rectangle, inside which is the name of the operation and other pertinent information about the operation. Arrows show direction of flow. Insert into each rectangle a symbol that represents your best estimate of: the probable type of contamination; likelihood of survival or destruction during heating or other processes designed to inactivate pathogens or toxic substances; or the likelihood of multiplication of pathogenic bacteria or toxigenic molds. Also measure or gather information about temperature and duration of the process, specify size of containers and depth of food in the containers, and note name of person performing the operation on Form H. An example of a flow diagram with noted information is illustrated in Fig. 1 (page 27).

Interpret Fig. 1 in the following manner: Raw frozen turkeys are likely to be contaminated (e.g., by *Campylobacter jejuni*, *C. perfringens*, *Salmonella*, *Staphylococcus aureus*, and *Yersinia enterocolitica*). Subsequent contamination could have come from persons who handled the product or equipment that contacted the product while it was prepared on a table, deboned, put into pans, or cut or ground. Cross contamination could have occurred from wiping a table soiled with thaw and drip water and then using the same cloth to wipe a cutting board or knife before deboning. Cross contamination could have also resulted from Worker B, who handled the raw turkeys and later handled the cooked turkeys. Without specific temperature measurements, it is unknown whether vegetative bacteria would have survived cooking, but many types of bacterial spores would survive. The time (3 h) is short, so vegetative bacteria may have survived, particularly if the turkeys were not completely thawed.

Bacterial growth could have occurred readily in the 6-in. (15 cm) depth of turkey meat and the 5-gal (19 L) container of stock during cooling. Vegetative cells in the turkey, however, would have quite likely survived the reheating and subsequent

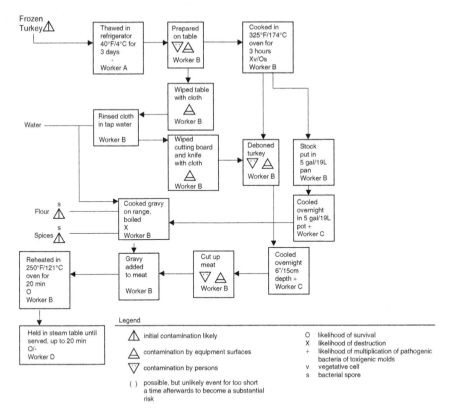

Fig. 1 Flow chart for preparation of 20-pound/9 kg turkey

holding in the steam table, but they would have been killed in the gravy that was boiled. The operations of concern should relate to those specified as likely contributory factors in the keys (A–F). During an investigation all such relevant information concerning the operation and food sources would be recorded on Forms H, I and J1-3.

As information is received and recorded on the flow chart, several hypotheses will be formed. Confirm the story by talking to persons involved or who observed the operation and, if practicable, by sampling and testing foods in question and by taking measurements at critical control points during preparation at a later time. Modify the flow chart as precise information is uncovered. For certain operations that call for more detailed evaluations, checksheets may be designed.

Review Monitoring Records

Review monitoring forms for date, time, and temperatures recorded, and persons doing the monitoring; entries of deviations from critical limits; notations of corrective actions taken whenever deviations were cited; and suggestions that the entries were falsified. The falsification may be characterized by the exact critical limit entered frequently, very similar daily entries, uniform entries, and illogical entries suggested by experience or knowledge of the operation and typical entries. Follow up on all suspicious entries for the dates and foods under investigation. Furthermore, observe the way workers monitor and correct deficiencies when food safety criteria are not met. Record appropriate information on Form H. Check calibration of the monitoring equipment used by establishment staff.

Interview Food Workers

Interview separately all persons (e.g., chef, processing line worker, storekeeper) who were directly involved in producing, harvesting, processing, packaging, preparing, or storing the food under investigation and other persons who could have observed these operations (e.g., plant operation supervisor, table servers, kitchen assistants, and cleaning staff). Be aware that potential cultural and language barriers can make interviews difficult. A different interviewer may be needed to accommodate these barriers. If appropriate, talk to managers and workers involved in producing, transporting, processing, packaging, preparing, or storing food at other levels of the food chain, as well as persons who prepare food at home. Also ask the questions listed in the "Workers" section.

Ask questions in a sequence to reveal the flow of food from time of receipt until served, taken out, delivered, or shipped. Ask each worker to describe and perhaps demonstrate the way operations were carried out. The descriptions and demonstrations may reveal whether the worker understands the food safety aspects of the job. Also, get the workers' accounts of the manner by which the food in question was prepared and treated before they had contact with the food and after it left their possession. Use all this information to revise the flow diagram (Form G).

Review recipes or formulae for ingredients that may have been the source of the contaminant (e.g., eggs for *Salmonella,* raw or undercooked ground meat for *E. coli* O157:H7). Review the operational manual for procedures that may have led to contamination, survival, or proliferation of likely etiologic agents.

In foodservice establishments especially, ask preparation staff about foods that were prepared several hours or a day or more before being served at the suspect meal, and whether there were any unusual or different circumstances or practices while the suspect foods were prepared.

Food workers who think they could be criticized or suffer punitive action because of their possible role in the outbreak do not always accurately describe or enter on monitoring forms the food preparation as it actually happened. Their descriptions and entries should be plausible and account for possible sources of contamination and suggest possibilities of survival and potential for growth of pathogenic microorganisms. If the descriptions or other data do not contain all the wanted information, reword questions and continue the inquiry. Seek confirmation of one person's story by talking to others who have knowledge of the food operation and by watching food preparation or processing practices. Be alert for inconsistencies in the accounts of different persons. Be persistent until logical accounts are obtained. Record appropriate information on Form H and modify the flow diagram on Form G as appropriate.

As the situations are revealed, compare them with those that are likely as shown in the keys (A–F). Some of the entries specify contamination by persons who handle the foods, whereas others represent mistreatment of foods by persons involved with food operations.

Conduct Food Processing/Preparation Review

The investigation may involve places where foods were produced, gathered, caught, harvested, processed, stored, and/or prepared, or the means by which animals or foodstuffs were transported. Wherever possible, and perhaps at multiple places, conduct food processing/preparation reviews to determine what specific factors contributed to the outbreak (see Keys A–F). This vital aspect of the investigation should indicate critical control points and suggest control measures and monitoring procedures to prevent a recurrence at the site and other occurrences throughout the food industry. Thoroughness in conducting food processing/preparation review cannot be overemphasized.

At farms, investigate (a) records or history of animal illness and visits by veterinarians, (b) changes in production practices, (c) feed sources, and (d) water quality for animal drinking, hygiene of workers, irrigation, and spraying on crops, (e) flooding or changes in watershed management. Also investigate various operations such as (a) pesticide application, (b) fertilizing practices, (c) irrigation practices, (d) milking, (e) animal holding before slaughter, (f) fish and shellfish harvesting, (g) product washing, (h) cleaning food-contact surfaces, (i) storage procedures, and (j) worker health, as applicable. If shellfish are under consideration, evaluate records of diggers or harvesters who acquired the suspicious shellfish to identify those with a history of illegal harvesting, history of re-laying, evidence of shellfish tag switching, or use of fraudulent tags. The growing area one word may be contaminated by run-off water after heavy rain.

At slaughterhouses, watch dehairing, defeathering, washing, eviscerating, deboning, cutting up, and chilling of carcasses for potential sources of contamination and out-of-compliance activities. At processing plants, evaluate heat processing, cooling, freezing, drying, fermenting, acidifying, smoking, packaging, storing, movement of material within the facility, and/or other appropriate operations. At transport and warehouse facilities, evaluate records of food sources and consignees, opportunities for contamination based on items previously shipped, holding temperatures, items stored or transported at the same time, and adequacy of refrigeration.

At foodservice establishments, retail stores, open-air markets, and homes, investigate food sources, receiving, storing, preparing, cooking, handling after cooking, hot holding, cooling, reheating, and serving foods. If applicable, investigate phases of transporting, delivery, storing, and retailing. In particular, obtain information about when the suspect food was prepared, the ingredients used, and the source of any significant ingredient. Determine which workers were involved in preparing the food under investigation and the operations they performed.

Consider operations in establishments that processed or stored the food before or after the phase of the food chain that is under investigation. (See Keys A–F for listings of situations that likely contributed to foodborne disease outbreaks. Evaluate whether any of them occurred).

Complete copies of Forms G and H for each suspected vehicle. The information includes sources of foods and ingredients, persons who prepared the items, procedures used, potential sources of contamination during preparation, and time–temperature conditions to which foods were exposed. Starting at the time ingredients arrived at the establishment under investigation, include all temperatures and duration of each while the food was stored, transported, prepared, cooked, heat-processed, held warm, chilled, and/or reheated, and also during the interval between serving and sampling. Pay particular attention to time–temperature recording devices and charts and associated records. Conduct tests that may reveal flow of raw fluid foods into heat-treated foods, and record this information on Form H.

Be prepared to spend time making these observations and performing tests. Based on these discussions, observations, and measurements, revise the initial flow chart on Form G. Use the reverse sides or additional sheets to note additional observations or record data, as applicable.

Observe Operations

Observe from start to finish operations involving the product under investigation. Initially, persons may be tense or uncomfortable when they are being watched. As time goes by or as they get busy, however, they usually process or prepare foods using their customary food processing/preparation practices because of time limitations, habits, available equipment, and established procedures that must be followed. Practices intentionally modified to impress evaluators are usually obvious and may even result in inferior products.

Determine the likelihood that incoming foods bring foodborne pathogens into the establishment under investigation. Evaluate the manner of storage to decide whether it is appropriate in relation to the food's properties and type of packaging. As raw products are processed or prepared, determine the possibility that contaminants on them are spread to workers' hands, gloves, and equipment surfaces. Follow the processing to see whether the contaminated hands or gloves touch other foods and whether other foods are processed on this equipment. Observe practices of replenishing foods on display (e.g., dumping additional foods on remaining foods). Consider the possibilities of spreading microorganisms by cloths and sponges that are used to clean rawfood areas. Furthermore, determine procedures used for thawing frozen foods and reconstituting dried products. Watch postheating operations to see whether the foods are likely to become contaminated by persons who handle them or by utensils or equipment. Also, determine whether any ready-to-eat foods were handled with bare hands and whether the persons handling the foods had recent illness. If so, determine the date of onset, signs, and symptoms, and whether a physician was seen or specimens were collected. Observe hygienic practices, including hand-washing, of persons who process or prepare foods. Observe whether quantities of potentially hazardous foods are being stored beyond the storage capacity of the facility. Evaluate the effectiveness of cleaning and sanitizing utensils and equipment by watching cleaning procedures and examining appearance of the items after cleaning. Observe all aspects of likely contributory factors that are shown in the Keys (A–F). Modify, if necessary, the rough flow diagram and data on Form G as a result of the observation. Record other pertinent information on Form H.

Measure Time–Temperature Exposures of Foods

Check calibration of temperature-measuring devices before using them in foods and calibrate them if necessary. Then, thoroughly clean and disinfect them.

- Disinfect thermocouples by inserting the sensors into a pan of boiling water for a few seconds.
- Immersing them into a solution of at least 50 mg/L (ppm) hypochlorite for 30 s.
- Dipping them into 95% ethyl alcohol and immediately flaming them; or holding them over a flame from either a torch or ball of alcohol-soaked cotton.
- If flamed after immersing into alcohol, repeat three times. Use disposable alcohol wipes if these materials are not available.
- Disinfect thermometers by inserting the bayonet or bulb into boiling water or into a tube containing at least 50 mg/L (ppm) solution of hypochlorite for at least 30 s. (In certain foodservice operations, it may be feasible to keep water boiling in a pan on a stove or range throughout the day of investigation for disinfecting thermometers or thermocouple probes).

Measure temperatures of foods with thermocouples or thermometers. (Thermocouples measure temperature at precise locations accurately and rapidly). Insert the sensing end of the thermocouple probe into the precise location where a measurement is wanted. (The approximate center of a food is frequently used, but under certain circumstances another internal location or the surface may be chosen). Use bayonet-type thermocouples of appropriate length to reach the point to be measured in internal regions of food. If practicable, insert most of the shaft of the probe into the product being examined. Use thermocouples with button ends or reflecting potentiometers (temperature-measuring gun) to measure temperatures at product surfaces. If a reflecting potentiometer is used, position it close to and toward the food surface being measured. Plug thermocouple leads into a potentiometer and take readings at appropriate intervals.

If bayonet-type thermometers are used, insert the point beyond the approximate center to measure the highest temperature of a food being cooled or the lowest temperature of a food being heated. Raise and lower the thermometer as necessary to locate this zone.

Measure time with watches, timers incorporated in the potentiometer, a data logger, or recording devices. Record measurements and the times temperatures were measured on Form H, or make graphs from the data.

Measure product temperatures during processing and storage and record time sequences of operations. Measure ambient temperatures if they may affect product temperatures. Measure the temperature that foods of concern attained during or at the completion of initial heat processing or reheating and during postheating temperature rise and fall until the temperature drops to 130°F (55°C). For foods cooked in retorts or pressure cookers, evaluate the operation of the retort, temperature and time of processing, venting procedure, and adequacy of sealing, rather than product temperature. Verify calibration of the establishment's time and temperature and other measuring devices.

Determine whether temperatures and holding times of foods are within a range in which bacteria can multiply, and if so, whether they are likely to multiply rapidly or slowly. Evaluate the rate at which foods cool during storage at room temperature and in refrigerators and other cooling devices. Observe and measure temperatures of foods under investigation that are stored near heat sources because they may warm to ideal bacterial incubation temperatures and possibly be maintained at these temperatures for long durations. Measure the dimensions of containers used to hold foods being cooled and depth of the food mass. Record these on Form H. From these measurements, estimate probable cooling rates and the potential for bacterial growth. Determine whether covers are used (which impede cooling but may prevent further contamination or moisture and odor transfer); whether containers are stacked on top or against each other (which also impedes cooling); the location of containers in refrigerators (which may influence cooling or cross contamination); and whether there is forced-air flow or other types of rapid cooling (e.g., water/ice baths).

Plot time and temperature measurements on Form I, *Graph of Time–Temperature Measurements*, or on graph paper. Note that these measurements can be downloaded in a tabular or graphical form from the instruments, if data loggers are used. Temperature guidelines are included on Form I. They are:

- 250°F/121°C, a common retort temperature value at which bacterial spores are killed in minutes.
- 165°F/74°C, a temperature value at which vegetative forms of pathogenic bacteria are killed in a few seconds.
- 130°F/54°C, a temperature value at which vegetative forms of pathogenic bacteria are killed in a few hours.
- 120°F/49°C, a temperature value at or below which some pathogenic bacteria can multiply.
- 70°F/21°C, an approximate temperature value at which the bacterial log phase significantly increases and the geometric growth rate begins to slow. Rapid bacterial multiplication can occur between this value and the one listed

above, particularly between 115°F/46°C (which is the optimal temperature for the growth of *C. perfringens*) and 86°F/30°C (which is the optimal temperature for the growth of *B. cereus*). Optimal temperatures for all other foodborne pathogenic bacteria also fall between these values.

- 41°F/5°C, a temperature value near that commonly recommended for the cold holding of foods.
- 32°F/0°C, a temperature value at which only a few pathogenic bacteria can still multiply over weeks of storage, though most pathogenic bacteria cease multiplication at temperatures above this value.

Label product and ambient temperature curves and give notations of operations performed and related observations at the times plotted. Interpret the data on the bases of the optimal growth temperatures for microorganisms of concern and temperature ranges at which they can multiply. Also, based on the highest temperatures reached and time–temperature exposures, interpret heating and cooling curves to determine whether the pathogens of concern could have survived heating processes, or resumed growth, if undercooked or contaminated afterward during holding or cooling. Record pertinent information on Form H.

Measure pH of Foods

Several types of pH electrodes can be used to measure the pH of foods within an establishment. Some electrodes are encased in bayonet shafts that can be inserted into foods. Others that are commonly used for testing pH of laboratory media have a flat end and can be placed on food surfaces. The conventional laboratory probe is made to test pH of liquids. If one of these is used, the food must be either liquid or grounded or blended with distilled water (pH 7) that has recently been boiled (to drive off CO_2) and then cooled. Attach the electrode to a pH meter. Calibrate the meter (as recommended by the manufacturer) with at least two standard buffers (e.g., pH 4.0, 7.0, or 10.0) and compensate for temperature, if the meter does not do it automatically, before each series of tests. Measure pH of product at room temperature whenever possible. Thoroughly clean and rinse the electrode with either boiled and cooled distilled water or pH 7.0 buffer between each measurement. Record measurements on Form H.

Measure Water Activity of Foods

Before each measurement, adjust the hygrometer (a_w meter) to the a_w value 0.11 or lower with a standard LiCl salt solution. To test a_w of a food, put a sample into a small plastic dish, and put it into the vapor-tight chamber of the meter. Temperature influences a_w; therefore, keep the chamber at a constant temperature. Avoid fluctuations that exceed 0.3°C. (A fan within the cabinet helps maintain uniform temperatures.) Temperature fluctuations can be minimized in the field if the sample in the chamber (with attached sensor) is kept in a styrofoam box. A temperature of 30°C is recommended because of the ease of maintaining it; if 25°C is chosen, a refrigerated system in the incubator or unit is necessary. Some instruments are designed so that the chamber is kept in a water bath or is refrigerated or heated to maintain a constant temperature.

Use a standard salt (e.g., $MgCl_2$, NaCl, KCl, K_2SO_4) or sulfuric acid (H_2SO_4) solution to calibrate hygrometers to specific a_w values according to manufacturers' instructions. (The equilibrium relative humidity [ERH] values for certain salts at 30°C are: $MgCl_2$, 32.44±0.14; NaCl, 75.09±0.11; KCl, 83.62±0.25; KNO_3, 92.31±0.60; K_2SO_4, 97.00±0.40.) Select one that has an a_w value close to that of the samples to be tested. Calibrate the instrument regularly (e.g., whenever the drift exceeds 2–3%) to ensure a high degree of accuracy; this may require monthly calibration.

Allow time for the sample to equilibrate with respect to temperature and a_w. (This may take from a few minutes to several hours, depending on the size of the chamber, equipment used, or the type of sample.) Read the a_w from the digital readout or the recorder plot, or determine it from a calibration curve, as appropriate for the meter being used. Run duplicate samples whenever feasible, and average them for greater accuracy. (Equilibrium is usually considered to be achieved when two consecutive readings differ by less than 0.01 a_w units for direct readout equipment, or when a plateau is reached on recording equipment). Record measurements on Form H (pages 148–150).

Identify Contributory Factors of Outbreaks

During the investigation, identify factors that contributed to contamination and survival of the etiologic agents and perhaps also to their growth or amplification. These factors have been taken from reviews of reported foodborne disease outbreaks and outbreaks cited in scientific literature and surveillance data. They are listed and defined below according to a classification based on contamination, survival, and proliferation. Any factors contributing to outbreaks identified during an investigation would be recorded on Form H.

Factors that Introduce or Otherwise Permit Contamination

Natural Toxin. A toxic substance found in a plant or animal, or in some parts therein. For example, certain species of mushrooms contain one or more toxins. Certain algae (mainly dinoflagellates and other plankton) produce toxins that can be accumulated in shellfish, e.g. saxitoxin, or fish, e.g., ciguatoxin. Traditional approaches to food preparation are often designed to rid a food (or reduce the levels) of its toxic component, e.g., soaking of lupini beans. Persons who lack experience in a particular cuisine may not be aware of the utility of those measures.

Poisonous Substance Intentionally Added. A poisonous substance deliberately added to a food in quantities sufficient to cause illness. Poisons added because of sabotage, terrorism, mischievous acts, and attempts to cause panic or blackmail a company fall into this category (page 99).

Poisonous or Physical Substance Accidentally or Incidentally Added. Although there is seldom a deliberate attempt to add poisonous chemicals to foods, these substances can reach the food from spillage, indiscriminate spraying, or accidental contamination with cleaning chemicals. The misreading of labels on containers resulting in either mistaking poisonous substances for foods or incorporating them into food mixtures falls into this category. Hard or sharp objects can get into foods from (a) lack of removal (e.g., seeds, bone chips), (b) presence in soil (e.g., stones), (c) breakage during preparation (e.g., glass fragments), and (d) deterioration of equipment (metal fragments).

Addition of Excessive Quantities of Ingredients that Under These Situations Are Toxic. An approved ingredient in a food can be accidentally added in excessive quantities so as to make the food unacceptable for consumption. Examples include too great an amount of nitrites in cured meat or excessive quantities of ginger powder in gingersnaps.

Toxic Container or Pipelines. The container or pipe that had held or conveyed the implicated food is made of substances that are toxic. The toxic substance either migrates into the food or leaches into solution by contact with highly acidic foods or beverages. For example, a toxic metal (e.g., zinc-coated) container used to store highly acidic foods, or carbonated water that backflows into copper pipes fall into this category.

Raw Product or Ingredient Contaminated by Pathogens from Animal or Environment. The incoming animal to be processed or the carcass or cut of meat or poultry is contaminated with pathogens when it enters the processing or preparation operations. Examples are salmonellae and campylobacters on poultry carcasses. Because these occur frequently in low populations, this contributing factor is designated only when there has been laboratory confirmation of the same marker strain of the etiologic agent in the raw food or if a traceback identifies a flock or herd as the source.

Ingestion of Contaminated Raw Products. Contaminated products are ingested without being first subjected to any significant heat processing or cooking. Examples are raw milk and raw shellfish. This factor also includes foods that, for culinary purposes, are subjected to mild heat that is obviously insufficient to kill any pathogens present. An example is hollandaise sauce (containing raw egg yolk) that is mildly heated (i.e., heated to time–temperature exposures insufficient to kill vegetative forms of pathogenic bacteria or denature proteins).

Obtaining Foods from Polluted Sources. Foods obtained from sources shown to be contaminated (e.g., shellfish from sewage-polluted waters; crops recently fertilized with night soil, irrigated by sewage, or sprayed with polluted water; or harvesting fallen fruit from manure-fertilized fields).

Cross Contamination from Raw Ingredient of Animal Origin. Cross contamination can occur by one of several means. Raw foods or their fluids touch or drip onto foods that are not subsequently cooked. Foods not subsequently heated are processed on, or in, equipment that was previously used for raw foods of animal origin without intervening cleaning. Foods not subsequently heat-processed are handled by workers who previously handled raw foods without intervening hand washing. Equipment previously used for raw foods is cleaned with cloths, sponges, or other cleaning aids that were not cleaned or sanitized, including cutting boards. These cleaning materials are then used to wipe food-contact surfaces on equipment that will later process foods that are not subsequently heated. Equipment or utensils used for raw foods might have been used for ready-to-eat foods without first being washed, rinsed, and sanitized.

Bare-hand Contact by Food Worker. A food worker handles or otherwise touches with bare hands foods that are not subsequently cooked. This is a typical situation that precedes outbreaks caused by staphylococcal enterotoxins and enteric viruses (hepatitis A virus, norovirus and more rarely *Salmonella* Typhi).

Handling by an Intestinal Carrier of Enteric Pathogens. The carrier is colonized by the pathogen and does not effectively wash hands after defecation and touches the implicated food with bare hands.

Inadequate Cleaning of Processing or Preparation Equipment or Utensils. Equipment or utensils are either not cleaned between uses or the washing, rinsing, or sanitizing steps are insufficient to remove contaminants.

Storage in Contaminated Environment. This usually involves storage of dry foods in an environment in which contamination is likely from overhead drippage, flooding, back siphonage, aerosols or air flow, access by insects or rodents, and other situations conducive to such contamination.

Factors that Allow Survival of or Fail to Inactivate the *Contaminant*

Insufficient Time and/or Temperature During Cooking or Heat Processing. The time–temperature exposure during initial heat processing or cooking was inadequate to kill the pathogen under investigation. This does not include inactivation of preformed heat-stable toxins. In cooking, but not retorting, it refers to the destruction of vegetative forms of bacteria, viruses, and parasites, but not bacterial spores. If the food under investigation was retorted, then spore-forming bacteria would be included.

Insufficient Time and/or Temperature During Reheating. The time–temperature exposure during reheating or heat processing of a previously heated food (which often has been cooled, and frequently held overnight) was inadequate to kill the pathogen or inactivate heat-labile toxins. This does not include inactivation of preformed heat-stable toxins.

Inadequate Acidification. The quantity of highly acidic ingredients, the concentration of marinade, and/or the time of contact was insufficient to kill the pathogen of concern.

Insufficient Thawing Followed by Insufficient Cooking. Frozen foods were insufficiently thawed, resulting in survival of pathogens during cooking or heat processing.

Factors that Allow Proliferation of the Etiologic Agents

Allowing foods to remain at room or warm-outdoor temperature for several hours. When foods are left at room or warm-outdoor temperature for several hours, pathogenic bacteria multiply and proliferate to populations sufficient to cause illness, or toxigenic bacteria or molds elaborate toxins.

Slow Cooling. Foods are refrigerated in large quantities or stored in a manner in which rapid cooling is impeded, allowing pathogens to multiply. Several factors may influence the slow cooling, including large masses or volumes of foods in large containers; inadequate air circulation as can occur with tight-fitting lids; stacking of pans on top of others; and crowded storage. Storing foods in large containers is the most frequent example of this type of contributory factor.

Inadequate Cold-Holding Temperature. Refrigerators malfunction, their temperature is poorly controlled, or they have excessively high temperature settings. This is mainly a problem for cooked foods; raw foods usually spoil and will likely be discarded.

Note: The first three factors – allowing foods to remain at warm ambient temperatures, slow cooling during refrigeration, and inadequate cold-holding temperatures – can be combined into a single factor referred to as *inadequate refrigeration.*

Preparing Foods a Half Day or More Before Serving. If the interval between preparation and consumption is coupled with temperature abuse, there is ample time for pathogenic bacteria to proliferate to populations or produce toxins sufficient to overwhelm the resistance threshold of healthy adults. Shorter durations than a half day may lead to this consequence, but precise times of preparation and eating are not readily obtainable during investigations or from outbreak reports.

Prolonged Cold Storage for Several Weeks. This is a concern for psychrotrophic pathogenic bacteria (e.g., *Listeria monocytogenes,* nonproteolytic *Clostridium botulinum* types B, E, and F, *Y. enterocolitica*). Over sufficient durations, they multiply at ordinary refrigerator temperatures and proliferate to populations sufficient to cause illness or elaborate toxins (e.g., *C. botulinum* neurotoxins). This can be a concern also for production of histamine in certain fishes that have been stored for long durations.

Prolonged Time and/or Insufficient Temperature During Hot Holding. The temperature exposure of cooked foods that are held warm is insufficient to prevent bacterial growth and may even promote such growth over time. Typically this occurs during storage in steam tables, bains-marie, thermotainers, and other hot-air cabinets.

Insufficient Acidification. The concentration or quantity of highly acidic ingredients, the type of acid, or the duration of marinading was insufficient to prevent multiplication of pathogens in mildly acidified foods. Improper mixing of acid ingredients results in insufficient acidification of the food. Therefore, the food falls into the potentially hazardous food category even though it may be suspected, because of either its name or type of usual ingredients, to have a low pH.

Insufficiently Low Water Activity. The concentration of salts, sugars, or other humectants was insufficient to prevent multiplication of pathogens in foods that have not been refrigerated or have been inadequately refrigerated. Therefore, the food falls into the potentially hazardous food category even though it may be suspected to have a low water activity by its appearance or expected process.

Inadequate Thawing of Frozen Products. Foods are thawed in a way that is conducive to bacterial growth. The problem, however, is usually storage of foods at ambient temperatures for several hours or at refrigerated temperatures for several days, after the foods have thawed. When it is attempted to thaw foods in stacks of multiple units, the items on the outside frame will thaw before those in the center and reach temperatures conducive to bacterial growth while those in the interior are still frozen.

Anaerobic Packaging or Modified Atmosphere. These factors create conditions conducive to growth of anaerobic or facultative bacteria in foods held in hermetically sealed cans or in packages in which vacuums have been pulled or gases added. All anaerobic bacteria must have a low oxygen-reduction potential to initiate growth. While this condition was thought to be only found in foods put into a sealed package or container, anaerobic bacteria may also grow in food where there are pockets of anaerobiosis created as a result of the texture of the food or the manner in which it is stored.

Inadequate Fermentation. Starter culture failure or improper conditions to promote fermentation can result in a product (e.g., cheese, salami) containing pathogens or developing bacterial toxins.

In time, other factors will be identified. When this occurs, add them to the list. Evaluate the food and process to decide whether these events occurred. Check the appropriate boxes next to the listing of contributing factors on Form H.

Use Keys to Aid Detection of Contributory Factors

Keys A–F, Situations that likely contributed to outbreaks of foodborne diseases when the following were implicated as vehicles:

Meat or poultry [products]	80
Eggs, milk or milk products	82
Fish, shellfish, crustaceans, or marine mammals	84
Vegetables	88
Fruits, nuts, spices, grains, or mushrooms	90
Formulated or mixed foods	94

These provide information about many foods that have been implicated as vehicles of outbreaks and processed in various ways and for organisms likely to cause disease. Data in these tables are based on epidemiology, challenge testing, hazard analyses, and research on the ecology and toxicology of foodborne pathogens. Use the keys for guidance on likely sources and modes of contamination, likelihood of survival of heat processes, and opportunities for multiplication of microorganisms.

(Each entry of these categories is indicated in the subheadings by the initials *C* for contamination, *S* for survival, and *G* for growth, multiplication, or propagation). Typical vehicles are listed with subdivisions for different methods of processing. Contributory situations (which have been defined in *Identify Contributory Factors of Outbreaks* subsection) are given for raw products or ingredients before processing; processing or preparation operations that are likely to take place in food processing plants, foodservice establishments, or homes; and postprocessing or preparation abuse after foods are taken home or to social events. Hence, the situations progress from the source through consumption of the foods.

See the legend at the *top left corner* of each key for the meaning of the symbols. The most likely events, "Principal Factor to Consider," are indicated by an "✕." A ✓ represents a "Factor to Consider." A ▲ represents a "Potential Factor to Consider," a • indicates a possible "Source of contamination, but likely destroyed during processing," and a "T" indicates that a "Toxin Survives Heat Process." All of the situations are conditional depending on operations being performed and other circumstances. For example, the postprocessing/preparation outcomes depend on whether (a) the foods had been improperly processed, improperly handled, equipment improperly cleaned, (b) operations contributed to cross contamination, (c) a pathogen survived processing, (d) growth occurred after thawing, or (e) contamination occurred during or after rehydration.

Review the keys before entering a place under investigation. Once on site, evaluate situations or operations indicated by the symbols to help determine whether they contributed to the outbreak under investigation. Ensure that each of the process having symbols denoting different kinds of contributing factors are carefully evaluated, then document so that the epidemiologic database can be expanded.

Four examples demonstrate the way to use the keys. Turn to Key A (page 80) and follow the first example.

If raw or undercooked beef is suspected as being the vehicle in an outbreak (or a single case) of toxoplasmosis, first look down the agent column to *Toxoplasma gondii*. Likely contributing factors to the outbreak are an infected animal and the carcass and cuts of meat may have become contaminated during processing. Therefore, glance horizontally across the row to the • (Colonized/Infected/Toxigenic Animals) and the ▲ (Cross Contamination). A heat processing failure (indicated by ✕) during processing or at retail/home would allow any cysts to survive and potentially cause an infection (the risk of eating raw beef would be even greater).

Work through another example in the same key (A, page 80). However, this is a more complex situation with many possibilities for contributing factors. In an outbreak of enteritis (possibly caused by *C. perfringens*) in which roast beef is the suspected vehicle, note that most food animals will contain some level of *C. perfringens* spores in their gut. Thus, the source of *C. perfringens* is likely to be a colonized animal, its feces, and/or soil, grass or mud (✕). It is possible, but less likely, that contamination occurred from water or feed (▲). Other possible contributory factors occur when spores are transferred to the meat during slaughtering, further processing and preparation operations, including from external environmental sources and inadequate cleaning of equipment. This allows further spread to other products. Although it is unlikely that multiplication occurred in the raw product, insufficient carcass and meat product cooling can also promote some growth if the spores are allowed to germinate (all indicated by ▲). Spores of *C. perfringens* survive cooking and will germinate, and the resulting cells multiply, if the cooked beef is improperly held hot (too low a temperature), inadequate refrigeration temperatures, improper cooling by putting into refrigerators without reducing the volume, kept at room temperature for several hours, or inadequate reheating; since these are key factors allowing *C. perfringens* to grow to large numbers, they are all indicated by ✕ in the horizontal row assigned to the pathogen. Make a thorough review of all processing and preparation operations. Cross contamination from the environment at the place of processing or preparation including contamination during cooling, contamination by a food worker or someone on the household kitchen, improper cleaning of equipment, all may have occurred, but these factors are less likely than those previously stated. Thus, these operations are indicated by ▲. Nevertheless, evaluate all possibilities. If the roast beef is held at room temperature for several hours, or inadequately cooled overnight, even apparently normal reheating may be insufficient to kill the millions of vegetative cells that germinated from the spores or there were postcooking contaminants. Time–temperature abuse could occur when the roast beef was shipped from a processor or transported by a caterer. Evaluate these possibilities.

A third example concerns an outbreak of salmonellosis in which chocolate candy is implicated. (See Key F, page 94) Look down the product column to chocolate and the product category Formulated/Blended, and the agent column *Salmonella* and glance across the horizontal row. As shown by the symbols in the row, the source of salmonellae is likely to be either an infected animal (bird, rodent, reptile) or fecal matter that directly or indirectly contaminated one of the raw ingredients (cocoa bean, coconut, nuts [see Key E, page 92] or dry milk [see key B, page 82]). Roasting of the beans should be sufficient to kill any salmonellae present (thus, all these sources are indicated by ●), but if they survive or there is later cross contamination, any salmonellae present may well be viable in the final product, typically in low numbers but sufficient to cause an infection because of the protection by the high fat content of the chocolate. The organisms once present can survive in chocolate for many months and even years and cannot be eliminated by further processing without destroying the product. Thus, the ✕ goes for heat process failure. The use of contaminated water (⊗) refers to a leaking water jacket or pipe where the water is nonpotable. Additionally, the use of water in the process or condensation in the processing plant could have possibly allowed growth on surfaces, indicated under improper a_w adjustments by a ▲. Several other possible sources of contamination are shown by ▲. Buildup of moisture within the package, such as results from transfer from cold to warm storage or improper packaging and storage in moist environments, could also influence microbial stability. Evaluate these possibilities. Because mixed foods can have many components, some in small quantities, like spices, be sure to choose other appropriate keys for likely contaminants of individual ingredients when investigating formulated or mixed food.

If a food has not yet come under suspicion but a disease has, the Keys A–F (pages 80–97) can also be used to suggest a vehicle or possible contributory factors. For example, if *B. cereus* gastroenteritis is the disease, scan down the list in each key for this agent. Likely contributory factors include soil contamination, spore survival of heat processes, and growth during improper hot holding, improper cooling, inadequate refrigeration, and room/outdoor temperature holding. The emetic toxin would survive reheating and the vegetative cell would survive inadequate reheating. Evaluate these possibilities.

These examples serve to show how the keys give direction for identifying the most likely factors that led to the outbreak and other possible events that influenced contamination, survival of toxins or microorganisms, or multiplication of pathogenic bacteria or toxigenic molds. Common vehicles are listed in the keys. If, however, a food under suspicion is not listed, choose a similar food and investigate the indicated operations. If a formulated food is being investigated, check likely contributory factors for each ingredient in addition to those factors listed in Key F.

Furthermore, the keys can be used to indicate critical control points of food operations and to show factors that frequently contribute to outbreaks. These can be established by glancing vertically down columns to determine those with the most entries, particularly the solid entries. Priority for preventive measures is thereby indicated.

Chain of Custody Procedures for Collecting and Analyzing Suspect Foods in Foodborne Illness Investigations

Organize and Develop Chain of Custody Procedures for Sampling

In the world concerned about terrorism, sample collection in "routine" outbreak investigations takes on additional importance. Since it may not be evident that an intentional event has occurred until well into an investigation, all samples collected should be treated as potential evidence. All samples should be collected aseptically and a chain of custody maintained at all times. A discussion with law enforcement officials and additional training on collecting samples as evidence may be required. Regulations may specify how sample should be collected. If procedures are prepared in advance and used during "routine" outbreak investigations, samples appropriate for legal evidence will be obtained during the intentional event or during any legal action resulting from an unintentional contamination event.

Chain of custody procedures establish how samples are collected, shipped, received by the laboratory, prepared by the laboratory, and examined to obtain the final results. Sampling operations must be carried out using techniques that ensure the sample is representative of the suspect food(s) or lot(s) implicated in the foodborne investigation. Product samples must exhibit the same condition as before sampling, and the collection technique must not compromise the integrity and status of the suspect food or lot. The purpose of chain of custody procedures is to ensure that sample and investigative evidence collected during an epidemiological investigation is valid, maintained under proper control and document proper handling of samples so that analytical results can withstand scrutiny by the courts during legal proceedings resulting from the investigation. The validity of sampling procedures and sample handling are increasingly under scrutiny in legal cases resulting from foodborne illness investigations and deliberate food contamination.

An effective chain of custody system requires close cooperation between key public health regulatory personnel and legal advisers in your agency. When an agency contemplates developing or improving an existing chain of custody procedure, give top priority to working with the legal advisers in the agency. Then, identify a key person to create, coordinate, implement, and manage the system. This person must take responsibility to:

- Review the existing chain of custody system and procedures that could be incorporated into the foodborne illness investigation and sampling system
- Identify the information, steps, or procedures that cannot be incorporated from the existing system but that needs collecting or addressing in the updated procedures or system
- Identify ways to merge or integrate new procedures with the existing procedures or system
- Identify collaborating agencies and staff and other resource personnel who can assist and aid in chain of custody interpretations
- Train agency staff in the revised chain of custody procedures for sampling foods and environmental samples during a foodborne investigation
- Assemble sampling materials and forms that will be required during an illness investigation
- Periodically review and evaluate chain of custody procedures or system

Collect Samples of Suspect Foods

Using a menu or data from Table 5 or 6, decide which of the foods from the implicated meal are likely vehicles and take samples of them. Check storage areas for items that may have been overlooked. Collect ingredients or raw items used in the suspect food if they are likely sources of contamination. Also, check garbage for discarded foods or containers, because suspect foods may be thrown out if it is thought that someone may have become ill as a result of eating food produced or prepared in the establishment. Interpret these results with caution, however, because postincident contamination and/ or growth may have occurred, depending on the type of food, the ambient temperature, and the duration of the food in the container. If a food has not yet come under suspicion, collect samples of any potentially hazardous foods left from the suspect meal and of any foods available from an allegedly contaminated lot, as applicable to the situation. If deemed necessary, collect additional intact (unopened) packages or other containers bearing the same code number so that tests can be done for microorganisms, toxins, seam defects, vacuum, leaks, or other conditions, as appropriate.

If there are no foods left from the suspect meal or lot, try to get samples of items that have been prepared subsequently to the suspect lot but in a similar manner. When particularly hazardous situations are observed, collect samples of foods before and after heating, cooling, or other operations of concern.

Collect samples of foods aseptically and put them into sterile jars or sterile plastic bags. Use utensils (e.g., knives, spoons, tongs, spatulas) that are sterile and protected by a contamination-proof wrapper until used, or wash and disinfect utensils at the time of use. If disinfecting, use one of the following procedures:

(a) Insert into 95% ethyl alcohol and immediately flaming them
(b) Hold over a flaming alcohol-immersed ball of cotton
(c) Expose to a flame from a torch
(d) By inserting the utensil into boiling water
(e) By inserting the utensil into a container with at least 50 mg/L (ppm) solution of hypochlorite for at least 30 s; in certain situations, it may be feasible to keep water boiling in a pan on a stove or range throughout the day of investigation for disinfecting utensils

Provide the laboratory with a large enough sample unit for all the necessary examinations. (A sample unit weighing approximately 200 g or measuring approximately 200 mL will usually suffice. If only one analytical test is to be done, smaller portions may be collected). Coordinate sampling procedures with the laboratory before collecting samples.

Record temperature of the room, refrigerator, or warmer in which the food is stored just before collecting samples. Measure and record temperature of the food that remains after the sample unit has been collected. Optionally, if a plastic bag is used to hold the sample unit, squeeze the bag after filling to remove air. Wrap the filled bag around the sensing portion of the thermometer and hold it in place until the temperature stabilizes. If the sample unit is hot (>122°F/50°C), immerse it under running water or in a container of ice until the sample unit is cool to the touch. Take care to ensure that the sample does not become contaminated as a result of this procedure.

Label the container with an establishment identification code and sequential sample unit number. Keep a log of the code numbers, date, time of sampling, type of sample, and type of test to run. Complete Form F. Send a copy to the laboratory with the sample units and retain a copy for your files.

If sample units of perishable foods are not frozen at the time of collection, rapidly chill them to a temperature below 40°F/4.4°C. Keep them below this temperature until they can be examined. Do not freeze food samples because certain foodborne bacteria (such as gram-negative bacteria and vegetative forms of *C. perfringens*) die off rapidly during frozen storage. Pack sample units with ice or another refrigerant to maintain the desired temperature during transit. If wet ice is used there is a risk of the ice melting and contaminating the sample, so double bag both the ice and the sample. Transport sample units to the laboratory in an insulated container by the most rapid means. Keep them surrounded by, but not in direct contact with, the refrigerant (to prevent freezing the sample) until either logging and preparing analytical samples or transferring to a refrigerator at the laboratory.

Choose Appropriate Laboratory Tests

Choice of tests for the samples depends on supportive information that is wanted for the food processing/preparation review, type of foods being processed or prepared, and type of microorganisms or chemicals expected to be present. Aerobic mesophilic colony counts of food samples taken just after heating and again after holding give information about microbial growth during the holding period. They also can provide information on microbial inactivation (of common vegetative cells) if taken from raw materials and again after heating or from cooked foods after storage and again after reheating. These examinations have been shown to be quite useful and are within the capability of most technicians and laboratories. *E. coli,* coliform, thermo tolerant or fecal coliform, and Enterobacteriaceae are useful indicators of postheat processing contamination. *Salmonella* has been used to indicate survival of heat processing (e.g., egg pasteurization) or cross contamination of heat-processed foods. *E. coli* can also indicate cross contamination from meat and poultry. *S. aureus* can be used to indicate handling of heat-processed foods by human beings, as well as to provide evidence of risk of foodborne diseases. Enumeration of these microorganisms and aerobic mesophilic colonies can also indicate time–temperature abuse of food.

Specific foods may be tested for the presence or quantity of certain pathogens. For example, rice and other cereals, beans, milk, and potato dishes might be tested for *B. cereus;* fish and shellfish might be enumerated for *Vibrio parahaemolyticus;* cooked meat and cooked poultry products, gravies, and beans might be enumerated for *C. perfringens.* Epidemiologic information might suggest the need to examine certain foods for particular pathogens or indicator organisms. Data in Table B and the Keys A–F can serve as guidelines on tests to run for various foods.

Molecular techniques for DNA finger printing such as PFGE (pulsed-field gel electrophoresis) can provide helpful information to the investigation by identifying clusters of cases of illness and links among "sporadic" infections that are not geographically or temporally clustered. It can help establish a case definition by including or excluding cases based on their PFGE isolates patterns. In addition if the PFGE patterns of clinical and food specimens are the same, it is any more evidence for the casual association.

Recommend or Take Precautionary Control Actions

If there is strong evidence to form a hypothesis that the outbreak is foodborne, take applicable precautionary actions to stop the spread of the toxicant or pathogen. Avoid, however, over-interpreting data. Do not panic! The choice of action is dictated by:

- Known or suspected causal agent and the severity of its consequences
- Its source and type of vehicle involved
- Methods of food processing, packaging, and preparation to which the food has been subjected
- Distribution of the implicated food
- Availability of alternate eating sites or sources of food
- Treatment that the implicated foods are expected to receive before they are eaten
- Population at risk
- Cost of the possible actions in relationship to the risks of undesirable consequences
- Communications
- Administrative or political will to act

Rapid and appropriate actions are obviously warranted if the disease under investigation has severe manifestations (e.g., botulism), has a high probability of extensive spread of the agent (e.g., *Shigella*), or puts highly susceptible persons (e.g., aged population, infants) at risk. Take appropriate action (such as an embargo) to prevent distribution or serving of any suspect food until it is reprocessed, discarded, or proven safe. If the food is in intra- or interstate distribution, it is appropriate to discuss findings and conclusions with epidemiologic and regulatory authorities at state/province and/or national level before taking proposed actions. Previous investigations of outbreaks give precedents for the type of action to be considered.

Verify the effectiveness of these actions by monitoring illness incidence in the population to decide whether the outbreak terminates. If the high incidence continues, consider the possibility of other transmission routes. If processed foods are shown to be the vehicle or even if they are likely to be implicated in illness, warnings to the population at risk or recalls may be initiated. Suspend the precautionary action after adequate corrections are made that can be continuously monitored, or if during the investigation it is determined that the food or establishment under investigation is not involved.

Determine Source of Contamination

A complete epidemiologic investigation of foodborne illness includes an explanation of the source of contamination, as well as the manner by which the food became contaminated. The sources of etiologic agents are numerous (see Table B).

Compare Type of Isolates from Specimens with Those from Samples

If sources of contamination need to be identified or traced back to their source, definitive subtyping is usually essential. Definitive subtyping techniques that may be used for making associations between victim and food or between food and source of contamination are cited in Table 2 (page 46).

Identify Specific Source of Contamination

Physical substances often come from the soil from which plants are harvested; from bone, fin, and claw fragments; and from damaged or worn equipment. Chemical agents usually come from substances that are added to foods or that directly or indirectly reach food during production, processing, or preparation. Microbial agents have specific sources that include soil, water, products of animal origin, and human beings who handle foods (see Table B).

Raw Foods. Animals may be colonized with *Salmonella, C. jejuni, Y. enterocolitica, C. perfringens, S. aureus,* and/or other pathogens. Animal carcasses can become contaminated with these pathogens during slaughtering and processing, and the meats are often contaminated by the time they arrive in kitchens (see Keys A–B). If any of these agents are suspected in an outbreak, samples of meat and poultry, meat scraps, drippings on refrigerator floors, and deposits on saws or other equipment can sometimes be helpful in tracing the primary source of contamination. Swabbing

Table 2 Examples of definitive subtyping schemes for foodborne pathogens

Definitive typing scheme[a]	Pathogens
Antibiogram	*Vibrio* spp., *Aeromonas,* bacteria for which there are no other procedures
Colicin typing[b]	*Shigella sonnei*
Determination of invasiveness and enterotoxicogenesis[b]	Pathogenic *Escherichia coli, Vibrio cholerae*
Enzyme-linked immunosorbent assay (ELISA)	Hepatitis A virus
M and T typing	*Streptococcus* Group A
Molecular typing[b,c]	Research applications for all major pathogens, in addition to the listed pathogens for all of the following subgroups in specialized laboratories
Plasmid profile	*Campylobacter jejuni, Escherichia coli, Salmonella, Shigella*
Polymerase Chain Reaction (PCR)	Research applications in progress for detecting most types of foodborne pathogens
Pulsed-field Gel Electrophoresis (PFGE)	*Escherichia coli, Listeria monocytogenes, Salmonella, Shigella, Staphylococcus aureus, Vibrio cholerae*
Random Amplification of Polymorphic DNA (RAPD)	*Listeria monocytogenes, Vibrio cholerae*
Restriction Fragment Length Polymorphism (RFLP)	*Listeria monocytogenes, Escherichia coli* O157:H7
Multilocus Enzyme Electrophoresis (MLEE)[b,c]	*Listeria monocytogenes, Vibrio cholerae,* research applications for other pathogens
Mouse neutralization[a]	*Clostridium botulinum* types A-G types, A, B, E are most likely
Phage typing[a]	*Escherichia coli; Listeria monocytogenes; Salmonella* Typhi, *S.* Typhimurium, *S.* Enteritidis, and a few other serotypes; *Shigella; Staphylococcus aureus*
Serotyping	*Bacillus cereus, Campylobacter jejuni, Clostridium perfringens[d], Escherichia coli, Listeria monocylogenes, Salmonella, Shigella, Vibrio parahaemolyticus*
Toxin identification[b]	*Bacillus cereus, Clostridium perfringens, Staphylococcus aureus* types A–E
Shigatoxin/Verotoxin detection[b]	*Escherichia coli*

[a]New commercial kits are often becoming available following research and production
[b]Limited laboratory testing availability
[c]Potential for use for all organisms
[d]Previously available, and perhaps still available in some laboratories

surfaces of equipment (tables, cutting boards, grinders, and slicing machines) that contacted the suspect food and testing the swabs for pathogens of concern can sometimes establish links in the transmission of contaminants. When this is done, however, the isolates must be definitive typed (e.g., serotyping). This can be helpful to demonstrate the potential for cross contamination if a common utensil or piece of equipment is used for raw and then for cooked foods. Swab these surfaces with sterile swabs or sponges, moistened with a sterile solution (such as 0.1% peptone water or buffered distilled water). If using a swab, break off the tip of the swab into a tube containing 5–10 mL of this solution or into a tube of enrichment broth for a specific pathogen. If using a sponge, replace it into the original bag. Raw ingredients, whether of animal or plant origin, may be the initial source of pathogens. Therefore, identify which ingredients were added before and which were added after any thorough cooking or heat processing. Record this information on Form H (page 148).

Workers. Workers can be a source of foodborne pathogens. Enterotoxigenic strains of *S. aureus* are carried in the nasal passages of a large percentage of healthy persons. They are often found on the skin and sometimes in feces. *Clostridium perfringens* can be recovered from the feces of most healthy persons. Workers are sometimes infected with other enteric pathogens such as *Shigella, Salmonella* Typhi, hepatitis A and E viruses, and noroviruses, which are host-adapted to human beings. (See Keys A–F, and Table B, for data on sources of pathogens).

Ask food workers about their food consumption and illness history for a few days before and during the outbreak period. Evaluate whether their illness history meets the case definition. Check absenteeism by reviewing time cards and/or asking managers and supervisors. Determine the reason for any absence occurring a few days before and during the time of the outbreak. Was the absence due to a diarrheal illness? If so, compare the time of onset with the time of preparation of the epidemiologically implicated or suspect food or meal, and also with the incubation period associated with the disease under investigation. On Form C, complete information about signs and symptoms, a 3-day food consumption history, and other applicable parts for each worker who reported being ill during or before the outbreak. Record this information in the *Sources of Contamination* section of Form H: Be aware that workers may be victimized by their employers or coworkers if they are thought to have caused the outbreak. It is important that this part of the investigation is handled with great sensitivity.

Look for pimples, minor skin inflammation, boils, and infected cuts and burns on unclothed areas of the body; ask if there are any infections in covered sites. If deemed necessary, make arrangements for the workers to be examined by a physician.

If staphylococcal food poisoning is suspected, swab the lower half-inch of the nostrils of all persons who contacted the suspect food. Request a medical professional to obtain a culture from any skin lesion that is found. This is done by first removing surface exudate by wiping with 70% ethyl alcohol. If the abscess is open, aspirate, if possible, or pass two swabs into the lesion and firmly sample the lesion's advancing edge. (One swab is used for culture and one for gram stain.) If the lesion is closed, aspirate the abscess wall material with needle and syringe. Tissue or fluid material is always better than a swab specimen. Sampling of surface regions can introduce colonizing bacteria that are not involved in the infection process. Rectal swabs from food workers can be useful; for example, the responsible *S. aureus* may have come from the anus or perineum. Put each specimen in an individual tube containing a sterile preservative solution or transport medium, as recommended by laboratory personnel, and send them to the laboratory. If the disease under suspicion is norovirus gastroenteritis, hepatitis A, shigellosis, typhoid fever, or another disease for which human beings are the usual reservoir, get information about recent illnesses of all persons who may have touched the implicated food(s).

For bacterial infections, collect fecal specimens or rectal swabs from persons who handled the suspect vehicle. If norovirus is the suspected agent, obtain stool specimens from all workers who might have touched the implicated food. (Procedures are given in *Obtain Clinical Specimens* section, page 15). A less desirable alternative is to give each person who handled the suspect food a suitable container for a stool specimen. Then, instruct this person in its use, and state when the specimen will be picked up or describe the procedure for sending it to the laboratory. Be aware that persons who think that they may have been responsible for an outbreak may return a specimen from someone else. Other specimens may be needed, depending on the disease that is suspected (See Table B).

Complete Form E for each specimen. Take or send specimens with the report form to the laboratory.

If the same type of pathogenic microorganism is recovered from both a fecal specimen of a food worker and the suspect food, do not immediately conclude that the worker was the source. Consider the events that took place before the outbreak. A worker who ate some of the implicated food could be a victim rather than the source of the etiologic agent. A history that includes a skin infection (boil or carbuncle) or a gastrointestinal or respiratory disturbance preceding or during the preparation of the suspect food would be more incriminating.

Equipment. Evaluate the cleanliness and the manner and frequency with which equipment is cleaned. Seek opportunities for and possible routes of cross contamination between raw and cooked foods that have been processed or prepared on or by the same pieces of equipment. Collect samples from or swabs of, as applicable, air filters; drains; vacuum sweepings; food scrap piles; drawers in slicers, cutters or grinders; dried deposits on equipment; and dead ends of pipelines. These may reflect the presence of organisms that previously were in the establishment. Chill samples, unless of dry materials, and transport them as recommended (see *Collect Samples of Suspect Foods* section, page 42). Record information on Form H.

If injury has occurred because of biting on hard or sharp objects, examine equipment for missing fragments or parts, chips, or other signs of damage. Observe operations with the objective of determining ways in which the objects could have been introduced into the food(s).

Trace Contamination and Malpractices to Their Sources

If field investigation failed to detect the source of contamination (e.g., infected worker or contaminated equipment) at the place of preparation, contamination could have occurred before the food or ingredient arrived at this place. Therefore, trace the implicated food backwards through distribution channels to the place of origin. This may involve interstate/interprovince and international movement of the product to processing operations, sites of harvesting, farms, herds, or flocks. Each commodity has unique nuances and patterns of production and distribution. The distribution pattern may be complex because the food under investigation may have come from more than one supplier and each supplier may have distributed foods to multiple customers. Traceback investigations are done to identify distribution and sources of lots of foods that the ill persons ate.

Determine the lot involved from stock rotation practices and records, dates and quantities received, and dates and quantities used or prepared. Obtain the original shipping container or container label to determine supplier, manufacturer, grade, color, quantity, shipper, and both production and shelf life dates, if available and as appropriate. Also, obtain copies of invoices, bills of sale, bills of laden, air-freight bills, or receipts that document the supplier, dates of shipment and receipt, and quantities purchased. If the product requires special source labels, such as shellfish tags, obtain all tags received during the interval in question. Record information about the place of serving, place of preparation (if different), and supplier of suspect food or ingredient on Form J1 (page 152). Send these with the report, completed forms, and copies of invoices and other shipping documents to the next level of investigation, if appropriate.

Good traceback procedures require on-site contact with firms at each level of distribution. Obtaining the necessary information by telephone only without shipping documentation may give an erroneous traceback trail.

Develop a list of shipments and their dates that could be associated with the implicated food and lots from each step in the distribution chain. Record this information on Form J2 (page 153), *Food Traceback Report: Supplier to Sources of Implicated Food/Ingredient*. Include quantities and delivery times that the product was received and used on each date for each step in the distribution chain from the shipping records, stock rotation, and usage practices. Also, document outgoing quantities of products and shipping times to the next level of distribution. Use additional copies of this form as needed. Develop a timeline that is appropriate to the incubation of the disease under investigation and/or development of a parasite into the infectious stage for these shipments. Narrow the investigation to the suspected shipments most likely to have been involved. Relate these records and practices to the date of the suspected event. Be aware that more than one shipper may have been associated with the event.

Conduct food processing/preparation review at sites of processing, storage, harvesting, and production, as related to the origin and transport of the suspect foods. Identify possible modes of contamination and time–temperature abuse that allowed survival and/or proliferation during shipping, handling, or storage enroute; site of contamination; and any illegal activity involved (e.g., shellfish bootlegging). Record this information on Form H. If deemed necessary, collect samples at the different links of the distribution chain and test them for the etiologic agent. Record this information on Form M (page 160). Look for unique and identifying markers by comparing strains of isolates of the agent from the ill persons, the implicated food at the different places of distribution and use, and the original source (see Table 2).

The purpose of this information is to go to other establishments where the food was shipped to identify other persons at risk, whether they also became ill, and whether they ate food from the contaminated lot(s). Record information on Forms J1 and J2. Continue to enter information about earlier distribution channels and sources as the investigation unfolds. Some suppliers implicated during the traceback will try to counter suspicion by blaming the manner of contamination on the preparer, inadequate investigation methods, politics, or inaccurate traceback procedures. Therefore, provide them with epidemiologic data (e.g., food history attack rate table, statistical analysis) and laboratory results to show the validity of the investigation and suggest guidelines for preventive measures to gain cooperation.

Summarize traceback information by drawing a diagram on Form J3, *Flow Diagram of Product Source and Distribution,* page 154, with time notations that depict source, dates of shipments, dates of receiving the implicated lot, and dates of preparation and storage of the implicated food. The findings provide justification for recalls, embargoes of the contaminated food, and perhaps other control or preventive measures. Take the appropriate action to prevent further illness.

The process of tracing epidemiologically implicated food back through the distribution system to the place of processing or production may involve other community, state/provincial, or national agencies. Therefore, notify the appropriate agencies when traceback activity has begun and supply to them completed traceback reports and other appropriate documents to facilitate the traceback. The national agencies may provide traceback assistance, will verify traceback activities that have been conducted by state/provincial investigations, and will take over investigations when jurisdictions are exceeded. Update the flow diagram as traceback information is passed from agency to agency. Maintain open communications between the agencies involved.

Analyze Data

Organize and collate data obtained from the interviews of ill and well persons who partook of the suspect meal, who ate the suspect food, who attended a common event, or who were part of a family or group of which persons became ill. Summarize these data on Form D. Use appropriate calculations and analyses to:

- Classify the illness
- Identify affected groups
- Test the hypothesis as to whether the outbreak was associated with a common source
- Determine a vehicle
- Measure disease association
- Calculate confidence interval and statistical significance
- Determine the necessity for further field or laboratory investigation

Compare results of laboratory tests and observations made during on-site investigations and related calculations with epidemiologic data.

Plot an Epidemic Curve

An epidemic curve is a graph that depicts time distribution of onset of initial symptoms for all cases that are associated with the disease outbreak. (Each case is represented as a *small square*.) The unit of time used in construction of the graph depends on the interval covered by the outbreak, which will vary with the disease under investigation. For example, use a scale of days or weeks for diseases with a long incubation period, such as hepatitis A. Use a scale of hours, or groups of hours, for diseases with shorter incubation periods, such as salmonellosis. A rule of thumb is that the interval used on the *x*-axis should be no more than one quarter of the incubation period of the disease under investigation. Construct this graph using time of onset data from Forms C or D, employing the appropriate time scale. (Fig. 2 illustrates an example of an epidemic curve).

The epidemic curve helps to determine whether the outbreak originated from a common-source vehicle, such as water or food, or from person-to-person spread. A common-source epidemic curve is characterized by a sharp rise to a peak, followed by a fall usually being less abrupt than the rise, as shown in Fig. 2. The total range of the curve is approximately equal to the duration of one incubation period of the disease. By contrast, a propagated epidemic curve is characterized by a slow, progressive rise, and the curve continues over an interval equivalent to the duration of several incubation periods of the disease.

Secondary spread can result from person-to-person contact of initial cases with previously well persons. Include these cases in the epidemic curve. Secondary spread can occur in cases of outbreaks caused by norovirus, hepatitis A and E viruses, salmonellae, shigellae, and *E. coli*, for example.

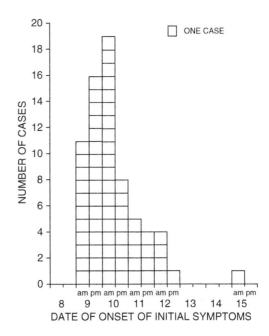

Fig. 2 Epidemic curve of a common-source outbreak

Table 3 Frequency of signs and symptoms

Signs and symptoms	Number of cases	Percent
Diarrhea	260	88
Abdominal cramps	122	41
Fever	116	39
Nausea	105	35
Headache	68	23
Muscular aches	56	19
Chills	55	19
Vomiting	42	14

Total number of cases = 296

Determine Predominant Signs and Symptoms

A symptom is felt by a person, whereas a sign is seen by an observer. Determine the percentage of ill persons who manifest each sign or symptom (as cited in Form C or D) by dividing the number of persons reporting the given sign or symptom by the number of cases (296 in the example in Table 3) and multiplying the quotient by 100. Determine which predominates by listing them in decreasing order of frequency and calculate the percentage of each as shown below in Table 3.

This information helps to determine whether the outbreak was caused by an agent that produced an intoxication, an enteric infection, or a generalized infection. Hence, it aids in classifying the illness into categories as presented in Table B. In the example given, the disease falls in the category *Lower gastrointestinal tract signs and symptoms (abdominal cramps, diarrhea) predominate.* (See Table B, and Form C). Such information is useful for formulating a case definition and making or verifying a differential diagnosis. Furthermore, it aids in choosing appropriate laboratory tests.

Determine the Responsible (or Suspect) Meal

When the approximate time of exposure for groups who ate common meals is not obvious, determine it by one of four methods.
(1) Calculate attack rates among the persons who ate different meals at which the contaminated food may have been served. To do this, divide the number of persons who became ill after they ate a particular meal by the number of persons who ate the meal, and multiply the quotient by 100. Do the same for persons who did not eat the particular meal. Make the calculations for all common meals that were eaten within the duration of the incubation period of the suspect illness. The responsible or suspect meal will be the one having the highest attack rate for those who ate the meal and the lowest for those who did not eat the meal. See Table 4, for an example. In this example, lunch on 1/17 has the highest rate for those who ate and the lowest rate for those who did not.

Table 4 Histories of meal attendance and attack rates

| Day/Meal | Ate/drank | | | | Did not eat/drink | | | | Difference in rates $a/(a+b)$ $-c/(c+d)$ | Relative risk[a] $\dfrac{a/(a+b)}{c/(c+d)}$ | p[b] value |
	Ill (a)	Well (b)	Total ($a+b$)	Attack rate (%) ($a/a+b \times 100$)	Ill (c)	Well (d)	Total ($c+d$)	Attack rate (%) ($c/c+d \times 100$)			
1/16B	52	100	152	34	58	94	152	38	−4	0.90	0.55
L	89	150	239	37	21	44	65	32	+5	1.15	0.55
D	87	150	237	37	23	44	67	34	+3	1.07	0.83
1/17B	56	105	161	35	54	89	143	38	−3	0.92	0.67
L	106	145	251	42	4	54	58	7	+35	6.1	<0.000001
D	78	130	208	38	32	64	96	33	+5	1.13	0.57

[a]See subsequent text for definitions and calculation procedures (page 59, 61)
[b]p value refers to statistical significance; see subsequent text (page 64)

(2) A similar approach is to compare percent difference and odds rations for case–control studies or relative risk for attack rate studies and the probability (*p*) of chance causing the difference. The results shown in Table 4 indicate that lunch on 1/17 is highly associated with illness and that the difference in rates would occur by chance less than 1 in a million times.

(3) If the disease itself has been identified or suspected, begin at the peak of a common-source epidemic curve (page 51) and count back the average duration (hours or days) of the incubation period of the disease. The suspect meal will fall within the incubation period range and should be close to the point of median incubation period.

(4) Make a graph which lists meals and dates eaten on one axis and each case on the other. Block in the squares that have been formed whenever a patient ate a meal at the suspected place. The suspect meal will be the one that the majority or all of the patients ate.

Calculate Incubation Periods and Their Median

An incubation period is the interval between ingestion of a contaminated food (containing sufficient quantities of pathogens or concentrations of toxins to cause illness) and appearance of the first sign or symptom of the illness. Calculate the incubation period for each case. Individual incubation periods will vary because of (a) individual resistance to disease, (b) differing amounts of food ingested, (c) high or low populations of infectious agents or concentrations of toxic substances in the food, and (d) uneven distribution of the infectious agent or the toxic substance throughout the food.

The interval between the shortest and longest incubation periods is the range. Determine the median incubation period, which is the mid-point value; half the incubation periods are shorter and half are longer. Therefore, it is the mid-value of a list of individual incubation periods that are ordered from the shortest to the longest. If the series comprises an even number of values, the median is the mean of the middle two values. The median, rather than the mean, is used because the median is not influenced by exceptionally short or long incubation periods that are sometimes reported in foodborne illness outbreaks, whereas the mean can be affected by aberrant values.

The median and range of the incubation periods, coupled with information regarding predominant signs and symptoms, form bases upon which to decide whether the illness in question is an infection or intoxication and also provides subgroups for classification into categories listed in Table B (page 109). This information suggests which laboratory tests should be performed.

Time of exposure may not be readily apparent because people usually eat several times each day. Therefore, the incubation period cannot always be determined for each case, and sometimes it cannot be determined for any. Careful interviewing, however, often uncovers one or more persons who ate the food in question at a specific time. Incubation periods often can be determined from such information.

Calculate Food-Specific Attack Rates (Retrospective Cohort Analysis) or Case–Control Exposure Percentages

Whenever a food served at a common meal or social event is suspect, prepare a food-specific attack rate table (e.g., Table 5) or a case–control exposure table (e.g., Table 6). The former is used for cohort studies when the entire group at the event is known and interviewed about illness and exposure. The case–control approach is used when all of those at risk cannot be identified or only a proportion of ill persons (cases) and well persons (controls) can be interviewed about their exposures.

Do Retrospective Cohort Analysis

The food-specific attack rate table compares the illness rate among those who ingested specific foods at an event or meal with the illness rate of those who were at the event or meal but did not ingest these items. To assist in this calculation, use Form K1 (page 155) and Table 5 as guides. To calculate the food-specific attack rate for a given item, divide the number of ill persons who ingested a particular food by the total number (both ill and well) who ate the food and multiply the quotient by 100. Do the same calculations for those persons who did not eat the particular food. An example in Table 5 is:

$$\frac{97 \text{ (ill who ate turkey; } a)}{133 \text{ (total who ate turkey; } a+b)} \times 100 = 73\%$$

Continue the calculation for each beverage or food for both those who ate the food and those who did not eat the food. Calculate the difference in rates for each food and enter the result in the column so named. Calculate a relative risk and p value for all foods that have a large positive percent difference. This provides better guidance in identifying the vehicle than percent difference.

   An attack rate table is useful for identifying the food that was most likely responsible for an outbreak. Identify the food in the table that has the highest attack rate for persons who ate it and the lowest attack rate for persons who did not eat it. The food with the greatest difference between these two rates is suspect. The suspicion is strengthened if this food was also ingested by the vast majority of persons affected. For example, in the above table, the attack rate for persons who ate turkey was 73%; the attack rate for persons who did not eat turkey was 8%. The difference in these two rates (percent difference) was +65%, which is greater than the difference for any of the foods listed. The relative risk shows that the attack rate for those who ate turkey was nine times greater (73/8) than for those who did not. The data also account for all of the 99 cases, most (77) of whom ate turkey. Turkey is, therefore, the suspect vehicle although there is also a high rate difference for dressing and to a lesser extent for peas and rolls. This is probably because those who ate the turkey also ate these other foods. Negative rate differences show no association.

Table 5 Food-specific attack rate table

| Food | Ate/drank | | | | Did not eat/drink | | | | Difference in rates $a/(a+b)$ − $c/(c+d)$ | Relative risk $^a(a/a+b)/(c/c+d)$ | p^b value |
	Ill (a)	Well (b)	Total	Attack rate (%) $(a/a+b \times 100)$	Ill (c)	Well (d)	Total	Attack rate (%) $(c/c+d \times 100)$			
Turkey	97	36	133	73	2	23	25	8	+65	9.1	<0.000001
Dressing	88	33	121	73	11	26	37	30	+43	2.5	0.00005
Peas	77	28	105	73	22	31	53	42	+31	1.8	0.0002
Rolls	50	16	66	76	49	43	92	53	+23	1.4	0.006
Pumpkin Pie	22	14	36	61	77	45	122	63	−2	1.0	0.9
Milk	12	6	18	67	87	53	140	62	+5	1.1	0.9
Coffee	59	39	98	60	40	20	60	67	−7	0.9	0.5

[a]See subsequent text for definitions and calculation procedures (page 62)
[b]p value refers to statistical significance; see subsequent text (page 57)

Some persons who did not ingest the suspect food or beverage, but who became ill, are sometimes tabulated as being ill. Plausible explanations are that some of these persons may have forgotten which food they ate; some may have become ill from other causes; a few may have exhibited psychosomatic rather than etiologic agent-induced symptoms; and if the etiologic agent was infectious in low populations, foods may have been cross contaminated. It is also not unusual for the table to include some persons who ingested contaminated food but did not become ill. Plausible explanations are that (a) organisms or toxins are not always evenly distributed in food and consequently some persons ingest small doses; (b) some persons eat smaller quantities than others; (c) some are more resistant to illness than others, and (d) some who became ill will not admit it.

Do Case–Control Studies

If the case–control exposure table is appropriate, use Form K2 (page 156) and Table 6, page 58 as guidelines. For both cases and controls, calculate the percentage of persons who ate a specific food and the percentage of persons who did not eat this food. Compare the two percentages and calculate an odds ratio and the *p* value. A *p*-value is defined as a measure of the chance that the observed results would occur if the null hypothesis were true (a null hypothesis typically corresponds to a general or default position). The probability associated with a statistical hypothesis will help decide if there is a significant association between exposure and illness or if the results are due to chance (coincidence). Usually, only a portion of those ill or at risk would be chosen for comparison because not all cases and controls can be identified or interviewed. Therefore, the numbers of cases and controls would usually be lower than those in the attack rate table. The same numbers, however, are used in this example to compare the two approaches. The odds ratios and *p* values provide higher confidence and better guidance for identifying the correct vehicle than do the percent differences, which is shown in this example.

Particular care must be taken in the selection of controls. Cases are persons who meet the case definition. Controls should not be everyone else at risk. For example, whenever laboratory-confirmed cases are used as the criterion for the case definition, persons whose symptoms are compatible with the syndromes (as listed in Table B) should not be included as a case and should be excluded from the control group.

The totals for cases (ill) and controls (well) are fixed and do not change with each food unless there were unknown responses as to whether certain foods were eaten. The example in Table 6 shows a difference of 37% for turkey. Rolls, pie, coffee, and milk can initially be excluded because the number ill does not approach the number ill for turkey. Dressing and peas need further analysis.

Do Stratified Analysis

In some outbreaks, stratified analysis can be used to compare foods that have similar attack rates. As illustrated in Table 7 (cross table), attack rates for those eating and those not eating one food (e.g., turkey) are compared with attack rates for those

Table 6 Case–control exposures

Food/Beverage	Cases (III)				Controls(Well)				Difference in percent $-b/(b+d)$	Odds ratio[a] (ad/bc)	p[b] value
	Ate (a)	Did not eat (c)	Total $(a+c)$	Percent exposed ×100	Ate (b)	Did not eat (d)	Total $(b+d)$	Percent exposed ×100			
Turkey	97	2	99	98	36	23	59	61	+37	31.0	<0.0000001
Dressing	88	11	99	89	33	26	59	56	+33	6.3	0.000006
Peas	77	22	99	78	28	31	59	47	+31	3.9	0.0002
Rolls	50	49	99	51	16	43	59	27	+24	2.7	0.007
Pumpkin Pie	22	77	99	22	14	45	59	24	−2	1.0	0.98
Milk	12	87	99	12	6	53	59	10	+2	1.2	0.91
Coffee	59	40	99	60	39	20	59	66	−6	0.9	0.5

[a]See subsequent text for definitions and calculation procedures (page 62)
[b]p value refers to statistical significance; see subsequent text (page 57)

Table 7 Stratified analysis comparing food-specific attack rates for eating and not eating two foods

		Ate	Did not eat dressing	Totals
Ate turkey	Ill	88	9	97
	Well	33	3	36
	Total	121	12	133
	Percent ill	73	75	73
Did not eat turkey	Ill	0	2	2
	Well	0	23	23
	Total	0	25	25
	Percent ill	0	8	8
Total	Ill	88	11	
	Well	33	26	
	Total	121	37	
	Percent ill	73	30	

eating and those not eating another food (e.g., dressing). The total values in the table correspond to values in the food-specific attack rate table, but the cells within the stratified analysis table must be obtained from data on Form D2 or from individual food histories (Form C2).

High attack rates resulted when turkey was eaten (73% and 75%) and low attack rates when it was not eaten (0% and 8%), whether or not dressing was eaten. This was not so for dressing (73% and 0% and 75% and 8%, respectively). This comparison provides additional evidence that turkey was the vehicle in the outbreak.

Calculate Food Preference Attack Rates

In situations where common meals were not eaten by the ill persons, when investigating a disease that has a long incubation period (e.g., hepatitis A, typhoid fever), or when there has been a long duration between onset of illness and interviewing, a food preference attack rate table may help to identify the food vehicle. An example is shown in Table 8. In this example, a higher attack rate is observed in the group who purchased and ate Brand X cheese; a low rate occurred in the group not

Table 8 Food preference attack rate table

Food	Always or usually eat (Purchased within incubation period)				Never eat (Not purchased within incubation period)				Percent difference
	Ill	Well	Total	Attack rate	Ill	Well	Total	Attack rate	
Milk, Brand A	17	116	133	12.8	5	20	25	20.0	−7.2
Milk, Brand B	9	85	94	9.6	13	51	64	20.3	−10.7
Cheese, Brand X	22	102	124	17.7	0	34	34	0.0	+17.7
Cheese, Brand Y	20	125	145	13.8	2	11	13	15.4	−1.6
Cheese, Brand Z	17	100	117	14.5	5	36	41	12.2	+2.3

eating it. Thus, Brand X cheese comes under suspicion as being the vehicle of the outbreak under investigation.

Make Statistical Calculations

To decide whether the observed association shows a causal relationship between exposure and disease, consider the following questions:

- How strong was the observed association between exposure and disease? Was it statistically significant?
- How consistent was the association between exposure and disease in this outbreak, and is it consistent with reports of other, similar outbreaks?
- How specific was the association between exposure and disease, i.e., did the same exposure always result in the same outcome?
- Was there a plausible time sequence, i.e., did exposure precede disease by a reasonable amount of time, considering the time of exposure and the incubation period?
- Was there a dose–response relationship? For example, were persons who consumed more food more likely to become ill?
- Is it biologically plausible that the suspected exposure caused the observed disease, so that all the data (including laboratory results from clinical specimens and food samples, epidemiologic observations, and on-site observations) fit together and make sense? There must be a rational explanation for contamination, survival, and proliferation.
- Was the same agent isolated both from persons who were ill and from the suspect food?
- Was there a consistent PFGE pattern among the cases?

Measurements of association reflect the strength of the relationship between an exposure and a disease and may be thought of as the "best guess" of the true degree of association in the population under investigation. The measurement itself, however, gives no indication of its reliability, i.e., the degree of credibility to assign it.

A test of significance gives an indication of the likelihood that the observed association is due to chance. The confidence interval is a range of values within which the mean lies, with 95% probability or confidence. The chi-square test statistic is influenced by both the observed magnitude of the difference and the size of the study, but it cannot distinguish the contribution of each. The measurement of association and the test of significance (or a confidence interval) provide complementary information.

There are sources of bias that are inherent to epidemiologic data collection and testing association between disease and exposure. These include selection, information, and confounding biases. An example of selection bias is when procedures used to select controls differ from those used to select cases. Cases identified often seek medical care, while less severely ill cases may not. Additionally, laboratory-confirmed cases represent only a portion of the total number of persons who became ill during an outbreak. Therefore, less severely ill cases may never become recognized. This results in an underestimate of the ill population.

Information bias may occur when retrospective questioning results in misclassification of case and control exposures due to poor recall. If misclassification is random, the result is usually an underestimate of the association; if misclassification is systematic, the association may be either underestimated or overestimated. Cases and controls may recall their experience differently. For example, cases who know or suspect that their illness is foodborne may recall eating a food that they did not actually eat or that they ate larger quantities than they actually did, whereas controls may not.

Confounding is caused by a second food or activity that is associated with illness and with the actual vehicle but is not actually causative. This bias can sometimes be corrected by calculating specific rates. Biases may be compounded by the mixing of large and small effects and may even change the direction of the effect.

Anticipate biases that may exist, and try to prevent or control them and estimate the direction in which they lead. Assess the situation to avoid over-interpreting or misinterpreting the data.

Measure Disease-Exposure Association

Two measurements of disease association – relative risk and odds ratio – are commonly used. The choice depends on the way data are analyzed. Relative risk is calculated from a cohort study, while odds ratio is calculated from a case–control study. Both calculations start with 2×2 contingency tables that compare ill and not-ill groups with exposure and nonexposure. An example is presented in Table 9. Both can be interpreted as "eaters have an x-times greater risk of illness than noneaters." Be aware of limitations of small sample sizes that result in imprecise measurements of disease and exposure.

Calculate Relative Risk (if applicable). The relative risk or risk ratio (RR) compares illness rates (number of ill persons/population) between populations known to have different exposure histories in a retrospective cohort study. It cannot be used in situations where the investigator chooses the proportion of ill and well persons to be interviewed. It can be used when all cases and the population exposed are known (i.e., cohort analysis). For example, it can be used for making associations of attack rates and for other situations of known populations.

Calculate the relative risks from data in Table 5 by using Form L1 and starting with data in Step 1 (see Table 9).

Table 9 2×2 table of data from Table 5

	Ill	Not ill	Total
Ate turkey	(a) 97	(b) 36	(a+b) 133
Did not eat turkey	(c) 2	(d) 23	(c+d) 25

$$RR = \frac{\text{Illness rates among exposed group}}{\text{Illness rates among unexposed group}} = \frac{a/(a+b)}{c/(c+d)}$$

$$RR = \frac{a/(a+b)}{c/(c+d)} = \frac{97/133}{2/25} = \frac{0.73}{0.08} = 9.1.$$

Interpretation:

$RR=1$: No difference in risk of illness between the exposed and unexposed groups.

$RR<1$: Exposed group has a lower risk of illness than unexposed group.

$RR>1$: Exposed group has a higher risk of illness than unexposed group.

In this comparison, those who ate turkey and were ill had a much higher (approximately nine times higher) illness rate than those who did not eat turkey. This shows a strong association between exposure and illness. It is not, however, proof of causality. This calculation assumes that other risk factors for those who ate turkey (exposed) and those who did not eat turkey (nonexposed) are approximately equal.

Calculate Odds Ratio (if applicable). The odds ratio (OR) is used in situations where it is not possible to obtain illness data on everyone who was exposed to a potential hazard. In such studies, the exposure histories of people with the illness (cases) are compared with the exposure histories of "similar" people (e.g., similar age, live in the same or similar neighborhood, attended the same event, and perhaps have other attributes in common) who did not become ill (controls). It is impossible to calculate true risk from a case–control study, but the odds ratio is used as an estimate of the risk.

$$OR = \frac{\text{Odds of exposure among cases}}{\text{Odds of exposure among controls}} = \frac{a/c}{b/d} = \frac{ad}{bc}$$

Using the 2×2 contingency table (see Table 9 and Form L1), fill in cells a, b, c, and d. Use data for turkey in Table 6 to calculate the odds that the people who became ill were more likely to have eaten the food in question than the people who did not become ill, as shown in Table 9. (Although the data comes from Table 6 rather than Table 5, the same numbers and totals, however, apply.)

$$OR = \frac{ad}{bc} = \frac{97 \times 23}{36 \times 2} = \frac{2,231}{72} = 30.9.$$

Interpretation:

$OR=1$: No difference in exposure between cases and controls; therefore the exposure being examined was not associated with the illness.

$OR<1$: The cases were less likely than controls to have been exposed to the suspect agent.

OR > 1: The cases were more likely to have been exposed to the suspect agent. Therefore, exposure may have contributed to illness, which is the situation in the example of data from Tables 6 and 9.

In this example, the odds of exposure were much greater for the cases than for the controls. Therefore, the odds of exposure (eating turkey) were greater for the group that was ill than for the group that was not ill, and turkey is likely to be the vehicle of the etiologic agent.

Another example is given in Table 10. In this situation, five cases of salmonellosis were reported and interviewed; three had eaten ice cream. Ten healthy persons were identified as controls; they had eaten at the same place.

Table 10 Outbreak table

Eating history	Case	Control	Total
Ate ice cream	3	3	6
Did not eat ice cream	2	7	9
Total	5	10	15

$$OR = \frac{ad}{bc} = \frac{3 \times 7}{3 \times 2} = 3.5.$$

In this example, the OR is greater than 1; this means that the ill group was more likely to have been exposed to the suspect agent than the control group. Therefore, exposure may have contributed to illness, but the numbers used for comparison are quite small and subject to error.

Evaluate Confidence Intervals for Measurements of Association

Both relative risk (RR) and odds ratio (OR) are point estimates of the true degree of association between exposure and disease. For each of these estimates, it is possible to calculate the plausible range of values for which there is a 95% chance that the range includes the "true" RR or OR. In other words, there is only a 5% probability that such results occurred by chance. The size of this confidence interval gives an indication of the precision of the point estimate, which is influenced by the sample size. If the confidence interval encompasses the value 1.0, that measurement of association (RR or OR) is not statistically significant; the observed association between illness and exposure could be due to chance.

Confidence intervals (CI) can be calculated by statistical computer packages (e.g., Epi Info™). For the previous examples, the 95% confidence intervals are:

Table 4:	RR = 6.1	95% CI = 2.4–15.9
Table 5/9:	RR = 9.1	95% CI = 2.4–34.6
Table 6/9:	OR = 31	95% CI = 6.9–278
Table 10:	OR = 3.5	95% CI = 0.23–59.1 (encompasses 1)

Interpretation: The 95% CI around the two risk ratios is much smaller than that around the odds ratios because the former is based on larger samples. In Tables 6 and 9, one cell is small, which influences the range of the 95% CI around the relative risk value. The odds ratio of 3.5 is not statistically significant because the 95% CI includes the value of 1.0 and the CI has quite a large range, which is related to the small numbers.

Test Statistical Significance

Statistical significance tests calculate the probability (*p*) that the differences in illness rates between those who ate the suspect food and those who did not eat this food were due to chance. Test relative risk or odds ratio values that exceed 1.0 to determine their statistical significance. Such testing can provide a degree of confidence that a particular exposure is the vehicle even without laboratory confirmation. Two tests, chi-square (χ^2) and Fisher's exact, are commonly used for this purpose. Computer programs (e.g., Epi Info™) are available for making such calculations, but a review of procedures for performing these calculations is given for those who do not have access to computers or appropriate software.

Initially a null hypothesis (H_0) – a hypothesis worded in a negative way – is stated. For example, a statement might be, "There is no difference in attack rates for those who ate turkey and for those who did not eat turkey." Or, "There is no association between eating turkey and getting ill." After statistical significance tests have been done, decide whether to accept or reject the null hypothesis.

Calculate chi-square (χ^2) *(If applicable)*. The chi-square (χ^2) test provides a means of comparing an observed illness distribution with an expected distribution. Observed distribution is the data used to calculate incidence rates such as food-specific attack rate. Enter case–control vehicle exposure percentage data on Forms L1 and L2 (pages 157–159), as applicable. Then, calculate the expected distribution. The expected distribution refers to a situation in which the attack rates would be expected to be the same among those who ate the suspect food and those who did not (indicating that illness was unrelated to food exposure). The level of confidence associated with obtaining different attack rates purely by chance (i.e., not attributable to any cause) must be chosen by the investigator. The most common confidence level used is that of 95%, which implies that chance alone would account for 5 out of 100 (1 in 20) sets of attack rates being statistically different.

Next, calculate the χ^2 statistic and determine the probability (*p* value), following the step-by-step procedure outlined in Form L1. Take the example of turkey from Table 5 or 6. Fill in cells *a*, *b*, *c*, and *d* from that table. Then fill in the marginal totals, *a+b* and *c+d*, from the table and record under substeps *i* and *ii*. Add *a+c* and *b+d* and fill in the other marginal tables and substeps *iii* and *iv*. Add the marginal totals (*a+b+c+d*) to obtain *n* (substep *v*). Therefore, the observed (outbreak) table would be as shown in Table 9.

Do calculations to complete the expected table (step 2, Form L1). To do this:

vi. Multiply *i×iii* (133×99) and then divide the product by *v* or 158 to get a_e = 83.3; round off quotient to the nearest whole number (83) and enter result into the a_c cell.

vii. Subtract a_e (vi) (83) from i (133) to get b_e = 50 and enter remainder into the b_e cell.
viii. Subtract a_e (vi) (83) from iii (99) to get c_e = 16 and enter remainder into the c_e cell.
ix. Subtract c_e (viii) (16) from ii (25) to get d_e = 9 and enter remainder into the d_e cell.

The expected table would be as shown in Table 11.

Table 11 Expected table calculated from data in Tables 5 and 6

Eating/drinking history	Ill/Case	Well/Control	Total
Ate turkey	83 (a_e)	50 (b_e)	133 ($a+b$)
Did not eat turkey	16 (c_e)	9 (d_e)	25 ($c+d$)
Total	99 ($a+c$)	59 ($b+d$)	158 (n)

The chi-square test is only accurate if all cells of the *expected table* are 5 or greater. Note that one cell in the *observed (outbreak) table* was 2 (which is less than 5), but all of the cells in the calculated expected table are 5 or greater. Therefore, the χ^2 test can be used to test the difference between the outbreak table and the expected table. (If any or all cells in the expected table are less than 5, skip steps 3 and 4 on Form L1 and proceed to step 5 on Form L2). Proceed with the χ^2 calculation as indicated on Form L1.

x. Multiply $a \times d$ ($97 \times 23 = 2{,}231$) and enter product on Form L1.
xi. Multiply $b \times c$ ($36 \times 2 = 72$) and enter product.
xii. Subtract xi from x ($2{,}231 - 72 = 2{,}159$) and enter the remainder; note that if xi is <0 the negative sign is dropped.
xiii. Divide n by 2 ($158/2 = 79$), and enter quotient rounded to a whole number.
xiv. Subtract xiii from xii ($2{,}159 - 79 = 2{,}080$) and enter remainder.
xv. Square xiv ($2{,}080^2 = 4{,}326{,}400$) and enter product.
xvi. Multiply xv by n ($4{,}326{,}400 \times 158 = 683{,}571{,}200$) and enter product.
xvii. Multiply i ($a+b$) xii ($c+d$) × iii ($a+c$) × iv ($b+d$) ($133 \times 25 \times 99 \times 59 = 19{,}421{,}325$) and enter product.
xviii. Divide xvi by xvii ($683{,}571{,}200/19{,}421{,}325 = 35.2$) and enter quotient, rounded to one decimal place. This is the χ^2 value.
xix. Convert χ^2 (35.2) to probability (see table in step 4, Form L1). Because 35.2 is greater than 7.88, the probability (p value) is <0.005, which is the smallest p value listed in this particular table.

The calculated χ^2 value of 35.2 has a probability (p value) of much less than 0.005. This is highly significant, as it indicates that the probability of the observed difference in the attack rates occurring by chance alone is much less than one time in 200 (or <5 times in 1,000). (In fact it is less than 1 in 10,000,000.) Therefore, it is highly probable that something other than chance was responsible for the difference between the observed attack rate and the expected attack rate. For instance, the turkey could have been contaminated by large populations of pathogens or a toxin and as a result could be the vehicle in the outbreak. Compare the results of the relative risk or odds ratio calculation with the χ^2 calculation of turkey data from Tables 5 or 6. The χ^2 calculation and associated p value confirms that the relative risk and odds ratio are statistically significant.

Calculate Fisher's Exact Probability (*If applicable*). The Fisher's exact test calculates the *p* value directly. It is particularly useful for calculations involving small numbers. There are two different ways to test the null hypothesis by the Fisher's exact test: a one-tailed test and a two-tailed test. The one-tailed test assumes a directional null hypothesis. The two-tailed test assumes a nondirectional null hypothesis, as in the χ^2 test. (Both are calculated by some statistical programs, e.g., Epi Info™).

In any situation in which one or more of the cell frequencies in the expected table is less than 5, or where any of the marginal totals in the outbreak table are less than 10, Fisher's exact test should replace the chi-square test. Fisher's exact test may be used to analyze any outbreak table. However, if the cell values of the outbreak table are all fairly large (i.e., >5), then the number of calculations required to determine the *p* value for Fisher's exact test becomes large. A high speed computer relieves this problem. Factorials are used in this calculation. [A factorial (!) is a number multiplied by all possible lesser whole numbers. For example, $5! = 5 \times 4 \times 3 \times 2 \times 1 = 120$. Note that $0! = 1$.]

An example of a one-tailed null hypothesis is, "The attack rate for those who ate ice cream is not higher than the attack rate for those who did not eat the ice cream." To calculate the one-tailed Fisher's exact test, the same 2×2 table (see Form L2) is used for Fisher's exact calculation as is used for χ^2 calculations (see Form L1). The outbreak table (data in Table 10, concerning cases alleged to be associated with eating ice cream) is given again but modified by the inclusion of attack rates for the example (Table 12). Calculate *p*1 from data in the 2×2 table. The *p* value for the observed distribution must be added to *p* values of those distributions that are more extreme in the same direction. Thus, develop all possible 2×2 tables that represent a more extreme result (i.e., greater difference in attack rate) than that of the outbreak table, but maintain the same marginal totals. This is done for *p*1.2 by decreasing the value in the *c* cell by one (i.e., $c-1$). *p*1.3 is the *p* value for the table created when $c-2$ is substituted for *c* in the outbreak table. Continue for *p*1.4, if necessary, and through *p*1.*x* until either *a*, *b*, *c*, or *d* = 0. When making manual calculations or using a hand-held calculator, all possible cancelations must be done, first for the factorials and then for the numbers resulting from converting the factorials to whole numbers before starting calculations. Tables 13 and 14 illustrate more extreme results. Observe that the totals ($a+b$, $c+d$, $a+c$, $b+d$, n) remain fixed while the "*c*" cell value is decreased by one. Each different cell frequency will have an associated 2×2 table, which will be more extreme than the previous one. If any of the cell frequencies is zero, a more extreme 2×2 table is not possible. The *p* value for the test become $pn = p1.1 + p1.2 + p1.3 + \ldots + p1.x$, where *p* 1.1 is the *p* value associated with the outbreak 2×2 table. This is necessary because the true *p* value is the probability associated with getting an attack rate difference of at least as large as that encountered in the outbreak table.

vi. To continue the example (see calculation approach in Form L2, steps *vi–x*):

$$p1.1 = \frac{6!9!5!10!}{15!3!3!2!7!}$$

Table 12 Outbreak table for case–control study associated with consumption of ice cream

Food consumption history	Case	Control	Total	(Attack rate)
Ate ice cream	3	3	$6(i_1)$	(50%)
Did not eat ice cream	2	7	$9(ii_1)$	(22%)
Total	$5(iii_1)$	$10(iv_1)$	$15(v_1)$	(33%)

vii. To simplify these mathematics, cancel factorial values, e.g.,

$$\frac{5!}{3!} = 5 \times 4.$$

Therefore, $p1.1 = \dfrac{6 \times 5 \times 4 \times 9 \times 8 \times 5 \times 4}{15 \times 14 \times 13 \times 12 \times 11 \times 2}.$

viii. To further simplify these mathematics, cancel whole numbers. Therefore,

$$p1.1 = \frac{3 \times 4 \times 5 \times 4}{7 \times 13 \times 11} = \frac{240}{1,001} = 0.240.$$

(*Note*: This p value (0.24) shows that, in 24 out of 100 times, chance could be responsible for the results; therefore, no further calculations need to be done for these data. The calculation, however, is carried on to show the procedures used).

Table 13 A more extreme table

Food consumption history	Case	Control	Total	(Attack rate)
Ate ice cream	4	2	$6(i_2)$	(67%)
Did not eat ice cream	1	8	$9(ii_2)$	(11%)
Total	$5(iii_2)$	$10(iv_2)$	$15(v_2)$	(33%)

Table 14 A more extreme table yet

Food consumption history	Case	Control	Total	(Attack rate)
Ate ice cream	5	1	$6(i_3)$	(83%)
Did not eat ice cream	0	9	$9(ii_3)$	(0%)
Total	$5(iii_3)$	$10(iv_3)$	$15(v_3)$	(33%)

$$p1.2 = \frac{6!9!5!10!}{15!4!2!1!8!} = \frac{6 \times 5 \times 4 \times 3 \times 9 \times 5}{15 \times 14 \times 13 \times 12 \times 11} = \frac{45}{1001} = 0.045$$

$$p1.3 = \frac{6!9!5!10!}{15!5!1!0!9!} = \frac{6 \times 5 \times 4 \times 3 \times 2}{15 \times 14 \times 13 \times 12 \times 11} = \frac{2}{1001} = 0.002.$$

ix. Calculate p (pn)-value for the one-tailed test

$$p = p1.1 + p1.2 + p1.3 = 0.240 + 0.045 + 0.002 = 0.287.$$

The p value from the above example for the one-tailed Fisher's exact test is 0.287, which is rounded off to 29 times in 100 that such results would occur by chance alone. A p value of less than 0.05 (5 times in 100 by chance alone) is necessary to consider the difference significant. In this example, although the cases were more likely to have eaten ice cream than the controls (as indicated by the odds ratio), this difference was not statistically significant. This result confirms the earlier finding that the 95% confidence interval of the odds ratio included the value 1.0, and thus the odds ratio was not statistically significant (see page 54).

In the two-tailed test, no assumption is made regarding which group (those who ate or did not eat ice cream) would have the higher attack rate. Therefore, the difference could be in either direction. For a two-tailed table, set up the same sort of tables and calculate p values for the worst case in the opposite direction as done in the one-tailed test. The $a+b$ (or i) is where the maximal number of the unexposed persons and the minimal number of the exposed persons were ill. (Therefore, the number in the 2×2 table's "a" cell may be, but is not necessarily, 0.) Adjust the values in all other cells, but maintain the same marginal totals. Continue increasing the number in the "a" cell by one in subsequent 2×2 tables and adjust the other values in the other cells. Do this until the difference in percents becomes less than the attack rate for the totals (33% in the example). Examples are shown in Tables 15 and 16. (*Note*: The one-tailed calculation for the p value, 0.24–0.29, shows that chance could be responsible for the results, and, therefore, no further calculations need to be made for this data. However, the calculation for the two-tailed test is carried on to show the procedures that would be used if the calculated p value was less than 0.05.)

Table 15 Table for data if no one was ill

Food consumption history	Case	Control	Total	(Attack rate)
Ate ice cream	0	6	$6(i_3)$	(0%)
Did not eat ice cream	5	4	$9(ii_3)$	(56%)
(percent difference)				(−56%)
Total	$5(iii_3)$	$10(iv_3)$	$15(v_3)$	(33%)

Table 16 Table for data if one person was ill

Food consumption history	Case	Control	Total	(Attack rate)
Ate ice cream	1	5	$6(i_3)$	(17%)
Did not eat ice cream	4	5	$9(ii_3)$	(44%)
(percent difference)				(−27%)
Total	$5(iii_3)$	$10(iv_3)$	$15(v_3)$	(33%)

Because the percent difference in Table 16 and all such tables with increasing number of ill persons is less than that of the totals, there is neither the need to go further nor to use this table in the calculation. Therefore, data from Table 15 is used to make calculations as above for $p1.4$.

$$p1.4 = \frac{6!9!5!10!}{15!0!6!5!4!} = \frac{9\times8\times7\times6\times5}{15\times14\times13\times12\times11} = \frac{6}{143} = 0.042.$$

Add this p value (or additional p values if data from multiple 2×2 tables are used) to the total for the one-tailed Fisher's exact test to get the p value for a two tailed Fisher's exact test.

xi. Calculate p (pn) value for the two tailed test = p value for one-tailed test + $p1.4$

$$p1 = 0.287 + 0.042(p1.4) = 0.329$$

The p value from the above example for the two-tailed Fisher's exact test is 0.329 (meaning that 33 times [rounded off] in 100 such results would occur by chance alone). No association is observed because a p value of less than 0.05 (5 times in 100) is necessary to consider the difference significant. Therefore, as stated above for interpretation of the one-tailed test, the attack rate for those who ate ice cream is not higher than the attack for those who did not eat the ice cream. This result confirms the earlier finding that the 95% confidence interval of the odds ratio included the value 1.0, and thus the odds ratio was not statistically significant.

These calculations of probability are based on a small number of people. If additional cases and controls could be interviewed, the difference in exposure may become more pronounced and possibly statistically significant; however, additional cases and controls also could provide more evidence that there is no statistical significance.

Interpret Results and Test Hypotheses

Record all laboratory results (e.g., specimens from cases, food samples, specimens from workers, environmental samples) on Form M, *Laboratory Results Summary* (page 160). Compare these results with epidemiological data and on-site observations. Use data obtained throughout the investigation to test hypotheses formulated during the investigation. Each of the following factors should be consistent with the suspected agent:

- Incubation period
- Type of illness
- Duration of illness
- Population affected
- Contributory factors leading to contamination of the food, survival of the pathogens from the effects of the process, and proliferation or concentration of the etiologic agent

Use applicable statistical tests to evaluate the significance of data generated during the investigation, but be aware of associated biases. The agent responsible for the outbreak can be determined by:

- Isolating and identifying pathogenic microorganisms from patients
- Identifying the same strain and PFGE pattern of pathogen in specimens from several patients
- Finding toxic substances, or substances indicative of pathological responses, in specimens
- Demonstrating increased antibody titer in sera from patients whose clinical features are consistent with those produced by the agent

The presence of some pathogens (e.g., *Salmonella, Shigella, E. coli*) in the epidemiologically implicated food is sufficient for confirmation. For other pathogens (e.g., *S. aureus, C. perfringens*), however, large numbers (e.g., 100,000/g or mL) must be recovered from foods. (See Table D). The large numbers indicate proliferation after contamination.

Sometimes laboratories test foods for indicator organisms (e.g., aerobic mesophilic colony count [standard plant count], coliform, fecal or thermotolerant coliforms, Enterobacteriaceae, staphylococci) rather than pathogens. The finding of these bacteria, even in high populations, in a food does not indicate that the food was a vehicle. High aerobic colony counts indicate that one of two situations occurred. The first is that the raw food or ingredients contained high populations of microorganisms and that the product received no or insufficient heat or other potentially lethal treatment to sufficiently decrease the population. The second is that the food was held at temperatures conducive to bacterial growth long enough to allow surviving spores to germinate and the resulting cells to multiply, or long enough to allow multiplication of contaminants that reached the product after a lethal process. Pathogens, if present, may or may not have multiplied along with the other flora; competitive flora may inhibit the growth of pathogens. The presence of soil-borne bacteria would be expected in foods grown in soil or exposed to it during growing and harvesting. Similarly, marine bacteria would be expected on raw seafoods. Fecal-associated bacteria (e.g., coliforms, fecal/thermotolerant coliforms, and Enterobacteriaceae) are likely to be found on raw foods of animal origin. Their presence in heated foods, however, suggests postprocessing contamination. Large populations of them suggest that they multiplied afterwards. Staphylococci on cooked foods indicate postcooking handling.

To confirm involvement of a suspect food, the same organism, toxin, or chemical markers must be found in the epidemiologically implicated food as were found in specimens from patients. The organism may be subtyped by serotype, phage type, immunoblotting, plasmid analysis, antibiotic resistance patterns, restriction endonuclease analysis, or nucleotide sequence analysis. (See Table 2) Even when clinical specimens are not available, a vehicle can be identified, at least circumstantially, by detecting toxic substances (such as zinc or botulinal toxin), by isolating a significant number of specific pathogens (such as 100,000/g or more *S. aureus* or *C. perfringens*) from the food, or by recovering enteric pathogens (such as *Salmonella*) from a food by enrichment techniques. The food from which these findings are made also should be epidemiologically suspect as a result of analysis of the food-specific attack rate table or case–control study, and the symptoms reported by the ill should be consistent with those produced by the agent that has been isolated from the implicated food. (See Table E).

Remember that testing samples can never replace observations, but test results can provide supportive data and often prove hypotheses. Take caution, however, when interpreting results of laboratory analyses. Negative results do not mean that pathogens are absent. Detection of the etiologic agent is more probable when the individual items or lots of food are heavily contaminated. Hence, a greater number of samples will be required for detection when the level of contamination is low. Counts follow probability distributions, which appear as a bell-shaped curve with the apex at the median, and therefore individual counts may have a considerable range.

Contamination of an individual food item is seldom homogeneous. To be homogeneous, solid foods must be ground or otherwise mixed thoroughly and liquids must be mixed. Only some portions of a particular food may be touched by a contaminated bare hand, or only some portions may contact a contaminated equipment surface. In time, the contamination is wiped, diluted, or removed from the contaminated surface. Therefore, the contaminants progressively diminish as the multiple items or surfaces of the same item continue to contact other surfaces. Many foods are solid, and spread of the contaminants throughout them is unlikely.

Furthermore, all items are not stored under identical circumstances, so propagation of bacteria may be likely in one or some items but not others, depending on storage conditions. Also, in any one item, multiplication of bacteria varies in different regions of the food. For example, bacterial growth may occur near the geometric center but not near the container surfaces of an item stored in a refrigerator because heat dissipates more rapidly at the surfaces exposed to cold air or surfaces. Growth of normal contaminating bacteria may overwhelm the pathogen, making its isolation difficult. Between the time the implicated food was eaten and the time samples were collected, pathogens may grow during storage to populations different from those ingested. Counts from such samples are difficult to interpret.

During exposures to heat or other processes that are detrimental to etiologic agents, time–temperature exposures vary in different portions of a food item. Therefore, survival of etiologic agents in the different regions of a food varies. Keep this information in mind when interpreting the laboratory results of a sample.

Unfortunately, the specific etiologic agent cannot be identified in many foodborne disease outbreaks. This is because food samples and clinical specimens have not been collected at an appropriate time, have been held too long, are too small a volume for effective analysis, or have not been examined for the appropriate agent.

The history of the way the food was produced, processed, prepared, or stored must reveal opportunities for contamination and, where applicable, for survival and growth of pathogens; otherwise, the history as recorded is incomplete or in error. If necessary, question food workers again and seek additional information, or look for inconsistencies in their stories that may suggest where contamination or other mishandling of the food occurred. The source of the causative agent can often be traced by recovering the agent from raw foods, food ingredients, equipment, food workers, or live animals or their environment. Definitive typing of isolates is required for confirmation (See Table 2). Such findings must also be supported by a history that would preclude the possibility of contamination from another source.

In practice, few investigations ever document and confirm all factors from the source of contamination to the onset of illness. Seldom is this even possible. Nevertheless, strive for the most complete investigation possible.

Make Recommendations for Control

After the food processing/preparation review has been completed, or as hypotheses are proved, take control actions or make recommendations to prevent further spread of the etiologic agent. These actions should be implemented in the place that the epidemiologically implicated food was eaten or purchased and other places to which a traceback investigation led. Record the action taken or recommended on Form N, *Control Actions Taken and Preventive Measures Recommended*. The following actions may be considered for the foods, personnel, or establishment involved.

Exclude Infected Persons from Handling Foods. Infected food workers are usually excluded from work when they show signs or describe symptoms of illness. Consecutive microbiological tests over an interval of several days or weeks may be necessary to demonstrate clearance of a pathogen, but despite this the pathogens may not be recovered during the examinations. Nevertheless, workers usually can return to work after recovery without such testing, if they practice good personal hygiene and are adequately supervised. If the illness appears to be caused by a highly infectious agent, e.g., *Shigella*, norovirus, take the following actions: (a) intensify hand-washing practices and (b) implement a no bare hand-contact policy for ready-to-eat foods. There are many sources of contamination other than workers, and workers who appear to be well may be infected without signs and may have practices as poor as or worse than those of the excluded worker(s).

Seize, Detain (Embargo), Stop Distribution, Remove, Recall, Reject, or Destroy the Epidemiologically Implicated Lot. Take appropriate action – depending on the type and degree of contamination and the estimated extent of the contamination – to stop the outbreak. For example, stop distribution and hold the food in locked facilities until tested and released or removed; reject the product at processing or preparation establishments or at ports of entry; remove the food from the premises and reprocess it under supervision; convert the food to animal feed; denature and bury, incinerate, or otherwise destroy the food; or recall all units of the implicated lots. For each product responsible for outbreaks, thoroughly evaluate their health hazard and mode of transmission before reprocessing or converting to animal feed. This is essential to ensure that foods contaminated with pathogens or toxins will not recur in the human food chain and not adversely affect animals. Many processors or caterers whose foods are under suspicion of being a vehicle voluntarily withdraw, and/or cease production and distribution of their products despite the associated financial losses and embarrassment.

Any product that has been deliberately contaminated may have to be destroyed though special means because of the potential presence of highly infectious or virulent organisms, or deadly toxins.

Cease Processing or Preparation of the Epidemiologically Implicated Food. When a vehicle has been identified, cease processing or preparation until corrections are made to eliminate situations that contributed to the outbreak. Ensure that the processes or preparation steps are modified to avoid or minimize contamination; to kill pathogens or inactivate toxins; and to prevent or significantly slow the growth of pathogenic bacteria so that a recurrence is prevented. Control criteria must be established or followed and the process must be monitored with sufficient frequency to ensure prevention of the events that led to the outbreak. Consider implementing a hazard analysis critical point system (HACCP).

Close the Establishment. When imminent risks to health exist if the operations continue, or when contributing factors that cannot be corrected are continuing, close the establishment. Consider reopening only when the contributing factors are identified and corrected or when the operation is brought up to industry standard. Because of legal ramifications, consult with supervisors before taking this action, but the prime consideration must be protection of the public health. Most operators will cooperate if their establishment is identified as the place where epidemiologically implicated foods were processed or prepared, and they may voluntarily offer to close. There may be specific legal implications if food has been deliberately contaminated and criminal actions are likely to proceed. It may not be feasible, however, to close an operation if a food was prepared for persons residing in an institution, unless an alternate, safe source of catered food is readily available. In such situations, consideration might be given to altering the menu to eliminate high-risk foods or to altering the operation to ensure safety, such as cooking foods thoroughly and serving them promptly.

Inform the Public

If there is a public health threat, announce the outbreak in the mass media so that the public who purchased the implicated food can be alerted to take action to return it to the place of purchase or other designated location; heat or otherwise prepare it safely; seek medical consultation or treatment; obtain vaccinations; or take prophylactic drugs. Consult with supervisors and medical personnel before taking the last two actions. Record actions on Form N (page 161).

Provide only objective, factual information about the outbreak. Coordinate among the investigating agencies to assure that a consistent and accurate message is delivered. It is often preferable to have one spokesperson for all agencies. Do not release preliminary information that has not been confirmed. The person giving information about an outbreak should be well informed about the etiologic agent being investigated and prepared to deal with questions. If the health hazard warrants a public warning at the hypothesis stage, tell the public why emergency measures are being invoked and that subsequent information may be cause to modify the action. As the investigation proceeds and the etiologic agent is confirmed and contributory factors are identified, terminate any emergency measures, and give advice on specific control and preventive measures. Attempt to reach all segments of the population at risk; this may require communication in multiple languages. Route all news releases or statements to all persons involved in the investigation. In situations involving large outbreaks or highly virulent or toxigenic etiologic agents, set up an emergency hotline for the public to call to ask questions. This is likely to occur if there is an intentional contamination event where there is high publicity and public concern (page 99). Train staff to handle these calls in a consistent manner so that the advice is the same no matter who gives it. Faulty information derived from poorly tested hypotheses can lead to severe political, legal or economic consequences.

They may then be disseminated by the mass media with inappropriate interpretations of the public health significance. Furthermore, this information may be used as an unrealistic base for food programs or food regulations because of either misinterpretations or pressure from misinformed consumer–advocate groups.

Calculate Economic Impact of Disease Outbreaks

Use data on costs of outbreaks to persuade officials in municipalities, state/provincial governments, national agencies, and legislatures to improve food protection programs by both regulatory and educational efforts; justify foodborne disease efforts of health and food regulatory agencies; and support disease surveillance. Such cost data on different outbreaks can help determine total costs of foodborne disease for a community, state/province, or country.

Try to estimate costs of some outbreaks. In one sense, the expenses of one group (e.g., ill persons) are recouped by other groups (e.g., drug companies, temporary-help agencies) or the group itself later (e.g., settlement compensations). The standard practice in estimating costs of outbreaks, however, is to document all costs relative to the incidents. Any benefits accrued can be listed separately and subtracted from losses. Some foodborne outbreaks have proven to be very expensive where a large population is exposed to a contaminated food. Businesses and tourism may suffer because of the adverse publicity, and government policies have changed. Subsequent control action may require the implementation of new regulations and educational programs.

Fill in Form O (page 162), *Economic Evaluation of a Foodborne Disease Outbreak,* as completely as possible when costing an outbreak. Many costs are not easily available or applicable; however, reasonable estimates relating to specific patients are almost as good as costs when deriving an overall cost of an outbreak (e.g., typical bed-care costs recall figures.) Direct costs are the easiest to measure. The main categories are medical and hospital care; investigation (e.g., epidemiology and laboratory analysis); loss to the food establishment, processor, or business; and productivity losses of ill or infected persons or those caring for ill persons (i.e., usually money not received or work not done for income paid). Estimate income for ill persons. If the exact values are not known, reasonable estimates are still useful because most costs are subject to question. If a daily wage is known, use it. Otherwise, estimate income per day from typical annual salaries for the occupation groups affected divided by 250 (i.e., 5-day work week plus 10 holidays) or another appropriate number. Although an unpaid homemaker has worth, no lost productivity value is normally assigned to this. Get details on food losses from a representative of the food supplier or establishment operator or representatives of insurance companies or legal firms.

Indirect costs are difficult to determine and any figures are subject to question. Although it does not result in financial losses, loss of leisure time is often considered equivalent in value to loss of work time. For retired persons, it could be equivalent to their pension, although they still get their full pension.

If deaths occur, calculate the cost of a life. An average value for a statistical life is hard to determine and varies from economy to economy. Recent estimates in the U.S. range from $6–9.1 million. Young children and elderly persons are normally calculated as being worth less than persons of wage-earning age. Deaths, however, are seldom attributed entirely to foodborne disease agents; other underlying causes often contributed to these deaths. For example, if an elderly or debilitated person suffers from foodborne illness, the fatality may be only 50% attributable to this syndrome. If death certificates are available, review these to decide the primary cause of death. However, information on these is often incomplete, and estimates can be made without them.

Even if some information is lacking, complete as much of the form as possible. Rational estimates can substitute for unknown data as long as the qualification is mentioned.

Submit Report

Summarize investigative data in a narrative report. State the case definition and methods used to gather clinical and exposure histories (e.g., telephone interview, questionnaires). Describe situations that led to contamination of the food, survival of the etiologic agent, and proliferation or concentration of the agent up to the time of consumption. Include all events that contributed to the outbreak to guide control and preventive measures. State control actions taken or recommended, recommended preventive measures, and the effectiveness of these actions (Form N).

Compare your data with the listings in Tables B, D, and E before assigning the etiologic agent and the vehicle. Outbreak confirmation is based on time, place, and person associations; recovery of etiologic agents from specimens, cases, and samples of food; identification of sources and modes of contamination; means by which pathogens or toxic substances survived treatment or proliferated. All of these, however, might not be found in any one investigation.

Complete Form P, *Foodborne Illness Summary Report.* Attach the narrative and the epidemic curve. Also attach Forms D2, G, H, I, J, K, M, and N and other data that will provide supplemental information to reviewers. If this outbreak has been part of a multijurisdictional outbreak you will need to coordinate your report with those other agencies. Consider a debrief of relevant stakeholders so that lessons learnt may be shared.

Send this report through administrative channels to the appropriate agency responsible for foodborne disease surveillance at state/provincial and national levels. Make the final report as complete as possible, so that the agency can accurately interpret the results and develop a meaningful foodborne disease data bank. Such a complete report is useful in the event of litigation. In interest of continuing cooperation, give all participants in the investigation due credit and send each a copy of the report. Also, send copies of the report through administrative channels to agencies that have jurisdiction over the implicated food, initiated the alert, and participated in the investigation. Consider publishing investigative data of important outbreaks (e.g., large outbreaks; unusual outbreaks, etiologic agents, or vehicles; outbreaks with associated mortality; new agents) in peer-reviewed journals.

Those concerned with food protection and with public health should make every possible effort to ensure the complete investigation and reporting of foodborne diseases. Without reliable, complete information, the trends in foodborne disease, incidence, and causal factors of the disease are difficult to determine. Good surveillance is essential for detecting and evaluating new foodborne disease hazards and risks.

Use Outbreak Data for Prevention

The primary purposes of a foodborne disease investigation are to identify the cause, establish control measures, and take actions to prevent future illness. Control measures can be effected either at the time of the investigation or immediately afterward by identifying a contaminated or otherwise hazardous product and removing it from the market.

To decrease incidence of foodborne illness, it is necessary to identify causal factors, develop practicable preventive procedures, and communicate them to those who can put them into practice. Factors shown by experience to contribute frequently to outbreaks are cited on pages 37–40, in Table B, pages 109–126, and Keys A–F, pages 80–97. Use Form H to fill out any factors observed. Inform managers, employees, and homemakers (as applicable) of the circumstances that contribute to outbreaks, and instruct them in proper food processing, preparation, and storage procedures.

Survey establishments that process or prepare similar foods to see whether conditions that contribute to outbreaks of illness are widespread. If so, initiate an industry-wide training program. If education fails to achieve the desired results, take other actions (such as targeting high-risk establishments with more frequent evaluations, conducting hearings, seizing contaminated lots, and prosecuting violators) to correct hazardous operational procedures. After such actions are taken, periodically evaluate these establishments to verify whether faulty procedures have been corrected or reintroduced into the operation, and to verify that critical control points are under control and being monitored effectively. If not, take appropriate educational and corrective action.

Alert the public to hazardous conditions that can affect them and motivate them to become concerned about their food supply. Only then will they insist on wholesome, safe foods that are processed and prepared in sanitary establishments. Consider establishing a newsletter containing reports that describe outbreaks, contributing factors, and preventive measures. Distribute copies to physicians, hospitals, the food industry, and others interested in foodborne illness and its prevention.

Foodborne illnesses are preventable, but prevention requires that those in the food industry and in health and regulatory agencies be constantly vigilant to ensure that the hazards are understood, and questionable operating procedures are avoided. Therefore, implementing a hazard analysis critical control point system can provide high assurance of food safety. This system is based upon: detecting hazards from epidemiologic data and observations of operations; identifying critical control points; establishing control measures, criteria, and critical limits; monitoring the critical control points to ensure that the process is under control; and taking immediate action whenever the criteria are not met.

Keys

Key A Situtations that likely contributed to outbreaks of foodborne diseases when meat or poultry were

		Farm/Field — Contamination Issues								Processing — Contamination Issues					
Meat or Poultry		Colonized/Infected/Toxigenic Animals	Animal Feces/Manure	Feed	Sewage	Soil/Grass/Mud	Water	Worker	Inadequate/Improper Cooling	Cross Contamination	During Cooling	Environment	Improper Cleaning of Equipment	Manipulation /Spread	Worker
MEAT															
Cooked, Pasteurized, and Other Heat Processes	**Bacteria**														
	Bacillus anthracis	●				●							▲		
	Clostridium botulinum		●			●									
	Clostridium perfringens	×	×	▲		×	▲			▲	▲	▲	▲	▲	
	Escherichia coli STEC/VTEC	●	●	●		●	●		●	×				✓	×	▲
	Listeria monocytogenes	●	●	●		●	●		●	✓				×	▲
	Salmonella	●	●	●			●		●	×	▲		✓	✓	▲
	Staphylococcus aureus	●						●		▲			▲		×
	Yersinia enterocolitica	●	●				●		●	▲				▲	▲
	Parasite														
	Taenia spp.	●								▲					
	Toxoplasma gondii	●								▲					
	Trichinella spiralis	●								▲					
	Virus														
	Hepatitis A virus							●							×
	Norovirus							●					✓		×
Cured/Dried / Fermented/ Smoked	**Bacteria**														
	Escherichia coli STEC/VTEC	●	●			●				▲		▲	▲	▲	▲
	Salmonella	●	●	●			●		●	✓		▲	✓	▲	▲
	Listeria monocytogenes	●	●	●		●	●		●	✓				×	▲
	Staphylococcus aureus	●						●		▲			▲	▲	✓
Retorted	**Bacteria**														
	Clostridium botulinum		●			●				✓					
	Staphylococcus aureus	●						●							×
POULTRY															
Retorted	**Bacteria**														
	Clostridium botulinum		●			●				✓					
Heated	Campylobacter jejuni	●	●				●			×	×		▲		
	Clostridium perfringens	×	×	▲		×	▲			▲	▲	▲	▲	▲	
	Listeria monocytogenes	●	●	●		●	●		●	✓				×	▲
	Salmonella	●	●	●			●			×	▲		✓	✓	▲
	Staphylococcus aureus	●						●		▲			▲		×
Cured/ Smoked/ Dried	Staphylococcus aureus	●						●		▲			▲	▲	✓

× = Principal Factor to Consider
✓ = Factor to Consider
▲ = Potential Factor to Consider
● = Source of contamination, but likely to be destroyed during later processing
T = Toxin Survives Heat Processes

Improper Hot Holding	Inadequate Refrigeration	Prolonged Storage	Room/Outdoor Temperature Holding	Heat Process Failure	Improper Cooling	Improper pH Adjustment	Improper Water Activity (a_w)	Inadequate Reheating	Organism/Toxin Survives Process	Cross Contamination	During Reconstitution	Improper Cleaning of Equipment	Worker/Person	Improper Hot Holding	Inadequate Refrigeration	Prolonged Storage	Room/Outdoor Temperature Holding	Heat Process Failure	Improper Cooling	Inadequate Reheating	Organism/Toxin Survives Process
									×												×
		×		×	✓	✓			×						×		×		✓	✓	×
×	×			×		×			×	▲		▲	▲	×	×		×		×	×	×
×	×			×	×	×			×		✓	▲		×	×		×	×	×	×	
	×	✓	✓	×	✓					▲		▲			✓	×	▲	✓	▲	▲	
×	×			×	×	×			×		✓	▲		×	×		✓	×	×	×	
×	×			×	▲	×			×	▲		▲	×	×	×		×	▲	×		×
	×			×	×	✓				▲		▲			×		▲	×	✓	✓	
				×						▲								×			
				×						▲								×			
				×						▲								×			
				✓									×					×			
										▲			×								
				✓		✓	✓		×	▲	▲	▲	▲								▲
				✓		✓	✓		×	▲	▲	▲	▲								▲
	×		✓	✓	✓	✓	✓		×	▲		▲			×		▲		▲	▲	▲
				✓		✓	✓		×		▲	▲	▲							T	▲
				×					×												✓
					▲				✓												
				×					×												✓
				▲	×					×		▲					×				
×	×			×		×			×	▲		▲	▲	×	×		×		×	×	×
	×	✓	✓	×	✓					▲		▲			✓	×		×		▲	
×	×			×	×	×			×	×		✓	▲	×	×		✓	×	×	×	
×	×			▲	×				×	▲		▲	×	×	×		×	▲	×		×
				✓		✓	✓		✓		▲	▲	▲							T	▲

82 Key B Situations that likely contributed to outbreaks of foodborne diseases when eggs, milk, or milk

Legend:
- × = Principal Factor to Consider
- ✓ = Factor to Consider
- ▲ = Potential Factor to Consider
- • = Source of contamination, but likely to be destroyed during later processing
- T = Toxin Survives Heat Processes
- C = Cooling Water Source of Contamination

Eggs or Dairy	Colonized/Infected/Toxigenic Animals	Animal Feces/Manure	Feed	Soil/Grass/Mud	Water	Worker	Inadequate/Improper Cooling	Cross Contamination	During Cooling	Environment	Improper Cleaning of Equipment	Manipulation/Spread	Worker
EGGS													
Bacteria													
Raw or Heated Lightly — Salmonella	×	×	×		✓		▲	▲				✓	
Heated — Salmonella	•	•	•		•		▲	▲			▲	▲	
Heated — Staphylococcus aureus	•					•							▲
Heated — Streptococcus pyogenes	•					•							×
Dried — Salmonella	•	•	•			•	▲	×			✓	×	×
MILK / MILK PRODUCTS													
Bacteria													
Raw — Brucella	×	×				✓					▲	▲	
Raw — Campylobacter jejuni	×	×				✓							
Raw — Escherichia coli STEC/VTEC	×	×	×		×	✓					▲	▲	
Raw — Listeria monocytogenes	×	×		×			▲			▲	▲	▲	
Raw — Mycobacterium bovis/tuberculosis	×					✓							
Raw — Salmonella	×	×	×		×	✓					▲	▲	
Raw — Staphylococcus aureus	×				×	✓					▲		▲
Raw — Streptococcus pyogenes	×				×								×
Raw — Yersinia enterocolitica	×	×			×	✓							
Cooked, Pasteurized, and Other Heat Processes — Escherichia coli STEC/VTEC	•	•			•	✓		▲	▲		▲	▲	
Cooked — Listeria monocytogenes	•	•		•		✓		✓	▲	×	✓	▲	
Cooked — Salmonella	•	•	•		•	✓		✓	▲		✓	▲	
Cooked — Staphylococcus aureus	•					•	✓	▲	▲		✓	▲	✓
Cooked — Yersinia enterocolitica	•	•			•	✓		▲	▲		✓	▲	
Dried — Salmonella	•	•	•		•	✓		×		×	✓	▲	
Dried — Staphylococcus aureus	•					•	✓				▲	▲	×
Dried — Yersinia enterocolitica	•	•			•	✓		▲		▲	▲	▲	
CHEESE													
Bacteria													
Fermented — Brucella	•					✓							
Fermented — Clostridium botulinum	•	•		•									
Fermented — Escherichia coli O157:H7	•	•	•			•	✓	▲					
Fermented — Listeria monocytogenes	•	•		•		✓	✓			×	✓	✓	
Fermented — Salmonella	•	•	•		•	✓	✓						
Fermented — Staphylococcus aureus	•					•	✓						×
Toxin													
Histamine													
BUTTER													
Bacteria													
Whipped — Staphylococcus aureus	•					•	×						×
Whipped — Listeria monocytogenes	•	•		•				×		×	✓		
ICE CREAM													
Bacteria													
Frozen — Salmonella	•	•	•		•	✓		▲					
Frozen — Salmonella Typhi					•	✓							×
Frozen — Staphylococcus aureus	•					•	✓						×

| | | | | | | | | | | | | | | Retail Store/Food Service/Home | | | | | | |
| Holding/Storage | | | Processing | | | | | | | Contamination | | | | Holding/Storage | | | Processing | | | |
Improper Hot Holding	Inadequate Refrigeration	Room/Outdoor Temperature Holding	Heat Process Failure	Improper Cooling	Improper pH Adjustment	Improper Water Activity (a_w)	Inadequate Reheating	Organism/Toxin Survives Process	Starter Culture Failure	Cross Contamination	During Reconstitution	Improper Cleaning of Equipment	Worker/Person	Improper Hot Holding	Inadequate Refrigeration	Room/Outdoor Temperature Holding	Heat Process Failure	Improper Cooling	Inadequate Reheating	Organism/Toxin Survives Process
	✓							✗		▲		▲			▲	▲				✗
	▲	▲	✗	▲						▲		▲		✓	✓	✓	✗			
	▲		T	▲				T		▲		✗	▲	▲	▲	▲		▲	T	T
	✓		✗	▲									▲	✓	✓		✗	✓		
			✗				▲	✗		▲	▲	▲	▲		▲		✗			
	✓	▲													✗	✓				
	✓	▲		✗											✗	✓				
	✓	▲		✗											✓	✓				
	✓	▲		✗											✗	✗				
	✓			✗											✗	✗				
	✓	▲		✗											✗	▲				
	✓	▲		✗									▲		✗	▲				
	✓	▲		✗									▲		✗	▲				
	✓	▲		✗											✗					
	▲	▲	✗	▲											✗	✗				
	✗	▲	✗	✗											✗	✗				
	✗	▲	✗	✗									▲		✗	✗				
	✗	▲	✗	✗				T					▲		✗	✗				
	✗	▲	✗	✗											✗	✗				
	✓	▲	✗	▲				✓		▲	▲				✓					
			✗					T		▲		✓								
			✗					✗		▲										
	✓		✗		▲	▲		✗	✗						✓					
					▲	▲		✗												
	▲	▲	✗	▲	▲	▲		✗	✗						✓	▲				
	▲	▲	✗	▲	▲	▲				✓		✓			✓	▲		▲		
			✗		▲	▲		✗	✗	✓					✓					
	▲		✗		▲	▲		T	✗				▲		✓					
					▲	▲		✗	✗											
	▲	▲	✗	▲				T						▲		▲				
			✗										✓							
	▲		✗	▲						▲		▲	▲							
			✗	▲						▲		▲	✗							
			✗	✓				T		▲		▲	✗							

Legend

× = Principal Factor to Consider
✓ = Factor to Consider
▲ = Potential Factor to Consider
• = Source of contamination, but likely to be destroyed during later processing
T = Toxin Survives Heat Processes

Fish/ Seafood		Farm/Field — Contamination Issues										Processing — Contamination Issues					
		Colonized/Infected/Toxigenic Animals	Environment/Climate	Feed	Sewage	Soil/Grass/Mud	Storage	Water	Worker	Inadequate/Improper Cooling	Prolonged Storage	Cross Contamination	Environment	Improper Cleaning of Equipment	Manipulation/Spread	Use of Contaminated Water	Worker
FISH																	
Raw	**Bacteria**																
	Salmonella			▲	▲		▲	▲	▲	▲		✓	✓	✓	✓		▲
	Parasite																
	Various (such as *Anisakis/Pseudoterranova, Diphyllobothrium latum* (fish tapeworm))	×	▲														
Retorted	**Bacteria**																
	Clostridium botulinum	•	•			•											
	Toxin																
	Histamine									×							
Heated	**Bacteria**																
	Salmonella			•	•		•	•	•	•		✓	▲	▲	▲		▲
	Vibrio cholerae (O1/non O1)		•			•		•	•	•		▲	▲	▲	▲	▲	▲
	Vibrio parahaemolyticus		•			•				•		▲	▲	▲	▲		
	Toxin																
	Ciguatoxin	×	✓														
	Histamine									×							
Smoked	**Bacteria**																
	Clostridium botulinum	•	•			•											
	Listeria monocytogenes		•			•			•			×	×	×	▲		
	Salmonella				•			•	•	•		✓	▲	▲	▲		▲
	Staphylococcus aureus								•			▲					×
Dried	**Bacteria**																
	Salmonella				•			•	•	•		✓	✓	▲	▲		▲
	Staphylococcus aureus								•			▲					×
Salted	*Clostridium botulinum*	•	•			•											
	Listeria monocytogenes		•			•			•			×	×	×	▲		
	Staphylococcus aureus								•			▲					×
Fermented	*Clostridium botulinum*	•	•			•											
MOLLUSCS (Clams, Oysters, Mussels)																	
Raw	**Bacteria**																
	Salmonella	×	×			▲		▲	▲	✓		▲	▲	▲			
	Salmonella Typhi	×	×			×			×	✓							×
	Vibrio cholerae O1	×	×			×		×	✓	✓		▲					✓
	Vibrio cholerae non-O1	×	×					×	×	✓				▲			✓
	Vibrio parahaemolyticus	×	×					×	×	✓				▲			✓
	Vibrio vulnificus	×	×					×	✓								
	Toxin																
	Azaspiracids	×	✓														
	Diarrhetic shellfish poison	×	✓														
	Domoic acid	×	✓														
	Paralytic shellfish poison	×	✓														
	Virus																
	Hepatitis A virus				×			✓									✓
	Norovirus				×			✓						✓			✓

Column groups — Left: **Holding/Storage** (Improper Hot Holding, Inadequate Refrigeration, Prolonged storage, Room/Outdoor Temperature Holding) and **Processing** (Heat Process Failure, Improper Cooling, Improper pH Adjustment, Improper Water Activity (a_w), Inadequate Reheating, Organism/Toxin Survives Process, Improper/Defective Packaging). Right under **Retail Store/Food Service/Home**: **Contamination** (Cross Contamination, Improper Cleaning of Equipment, Worker/Person), **Holding/Storage** (Improper Hot Holding, Inadequate Refrigeration, Prolonged storage, Room/Outdoor Temperature Holding) and **Processing** (Heat Process Failure, Improper Cooling, Inadequate Reheating, Organism/Toxin Survives Process).

Impr. Hot Hold.	Inadeq. Refrig.	Prolonged storage	Room/Outdoor Temp Hold.	Heat Proc. Failure	Impr. Cooling	Impr. pH Adj.	Impr. Water Act. (a_w)	Inadeq. Reheat.	Org./Toxin Survives	Impr./Def. Pack.	Cross Contam.	Impr. Clean. Equip.	Worker/Person	Impr. Hot Hold.	Inadeq. Refrig.	Prolonged storage	Room/Outdoor Temp Hold.	Heat Proc. Failure	Impr. Cooling	Inadeq. Reheat.	Org./Toxin Survives
	▲				▲						✓				✓		▲				
									✓												✓
			✗				✓		✗												
	✗								T						✗						T
	✓		✓	✗	✓				▲		✓	▲	▲	▲	✓		✓	✗	✓	▲	
	▲		✓	✗	✓						▲			▲	✓		✓	✗	✓		
	✓		✓	✗	✓						▲			▲	✓		✓	✗	✓		
									T												T
	✓		✓						T						✓		✓				T
	✓		✓	✗	✗		✓		✗	✗					▲		▲	▲	▲		▲
		✓		✗			▲		✗		▲	▲			▲	✗		▲			▲
	✓		✓	✗	✓		✓		▲		✓	▲	▲		▲		▲	▲	▲		▲
	✓		✓	✗	✓		✓		✗T		▲		✗		▲		▲	▲	▲		▲
							▲		✗		▲	▲	▲								▲
							✓		✗T			▲									▲
							✗		✗T	▲											▲
		✓					✗		✗	▲	▲	▲				✗					▲
							✓		✗T			▲									▲
						▲	▲		T	▲											▲
	✓		▲		▲						▲	▲	▲		✓		✓				
	✓		▲										✗		▲		✓				
	✓		▲		▲								✓		✓		✓				
	✗		▲		▲						✓				✓		✓				
	✗		▲		✗						✓				✗		✗				
	▲		▲												▲		▲				
													✓								
													✗								

(continued)

Fish/ Seafood			Farm/Field — Contamination Issues										Processing — Contamination Issues					
✗ = Principal Factor to Consider ✓ = Factor to Consider ▲ = Potential Factor to Consider ● = Source of contamination, but likely to be destroyed during later processing T = Toxin Survives Heat Processes		Colonized/Infected/	Toxigenic Animals	Environment/ Climate	Feed	Sewage	Soil/Grass/Mud	Storage	Water	Worker	Inadequate/Improper Cooling	Prolonged Storage	Cross Contamination	Environment	Improper Cleaning of Equipment	Manipulation/Spread	Use of Contaminated Water	Worker
Heated /	**Bacteria**																	
Smoked	*Salmonella*	●	●		●								✓	✓				▲
	Staphylococcus aureus									●	●		▲		▲			✗
	Vibrio parahaemolyticus	●	●						●		●		✓		▲			
	Listeria monocytogenes		●				●		●				✓	✓	✓	▲		
	Toxin																	
	Azaspiracids	✗	✓															
	Diarrhetic Shellfish Poison	✗	✓															
	Domoic Acid	✗	✓															
	Paralytic Shellfish Poison	✗	✓															
CRUSTACEANS (Crabs, Shrimp, Crayfish, Mudbugs)																		
Heated/	**Bacteria**																	
Smoked	*Salmonella*		●		●						●		✓	✓				✓
	Vibrio parahaemolyticus		●						●		●		✓		▲			
MARINE MAMMALS																		
Fermented	**Bacteria**																	
	Clostridium botulinum					●					●							

| | | | | | | | | | | | Retail Store/Food Service/Home | | | | | | | | | | |
| Holding/Storage | | | | Processing | | | | | | | Contamination | | | Holding/Storage | | | | Processing | | | |
Improper Hot Holding	Inadequate Refrigeration	Prolonged storage	Room/Outdoor Temperature Holding	Heat Process Failure	Improper Cooling	Improper pH Adjustment	Improper Water Activity (a_w)	Inadequate Reheating	Organism/Toxin Survives Process	Improper/Defective Packaging	Cross Contamination	Improper Cleaning of Equipment	Worker/Person	Improper Hot Holding	Inadequate Refrigeration	Prolonged storage	Room/Outdoor Temperature Holding	Heat Process Failure	Improper Cooling	Inadequate Reheating	Organism/Toxin Survives Process
	✓		✓	✗	✓						▲	▲		▲	✓		✓	✗	✓	✓	
	✗		✓	✗	✓				✗T		▲	▲	✗	▲	✗		✗	✗	✗	T	✗T
	✗		✗	✗	✗						✓	▲		▲	✗		✗	✗	✗	✗	
		✗		✗			✗		✗		▲	▲				✗		✗			
									T												T
									T												T
									T												T
									T												T
	✓			✗					▲		▲		▲		✓			✗			▲
	✗		✗	✗	✓						✓	▲		▲	✗		✗	✗	✓	✓	
						✓			✗T												✗T

Key D Situtations that likely contributed to outbreaks of foodborne diseases when vegetables were

Legend:

X = Principal Factor to Consider
✓ = Factor to Consider
▲ = Potential Factor to Consider
● = Source of Contamination, but likely to be destroyed during later processing
T = Toxin Survives Heat Processes

Vegetables	Farm/Field — Contamination Issues								Processing — Contamination Issues						
	Colonized/Infected/Toxigenic Animals	Environment/Climate	Animal Feces/Manure	Sewage	Soil/Grass/Mud	Water	Worker	Prolonged Storage	Cross Contamination	During Cooling	Environment	Improper Cleaning of Equipment	Manipulation/Spread	Use of Contaminated Water	Worker
HERBS/ GREEN ONIONS/PEPPERS (hot and mild)															
Raw / Dried — **Bacteria**															
Escherichia coli O157:H7	X	✓	X		▲	X			✓		✓	✓		▲	
Salmonella	X	X	✓	✓	▲	X			✓		✓	✓			
Shigella				✓		✓	X							✓	X
Parasite															
Cyclospora cayetanensis			X		X	X									X
Virus															
Hepatitis A Virus			✓		X	X									X
LEAFY GREENS															
Raw — **Bacteria**															
Escherichia coli STEC\VTEC	X	X	X		✓	X			✓		✓	▲		▲	
Listeria monocytogenes	X		X		X			X				✓		▲	
Salmonella	X	X	X	▲	✓				✓		✓	▲		▲	
Shigella				X		✓	X							▲	X
Parasite															
Various (such as *Cryptosporidium* and *Giardia*)	X		X		X	▲	✓		✓		✓			✓	✓
Virus															
Hepatitis A Virus			X			✓	X							▲	X
Norovirus			X			✓	X					✓		▲	X
POTATOES															
Heated — **Bacteria**															
Clostridium botulinum			●		●										
SPROUTS															
Raw — **Bacteria**															
Escherichia coli O157:H7	X	✓	X		✓	▲			✓		✓	▲	✓	▲	
Listeria monocytogenes					▲	▲			X		X	✓	✓		
Salmonella	X		X	▲	✓	▲			✓		✓	▲	✓	▲	
TOMATOES															
Raw — **Bacteria**															
Salmonella	X	X	✓			X			✓	✓	✓	▲	✓	▲	X
Shigella				▲		✓									X
VEGETABLES (all)															
Raw — **Bacteria**															
Escherichia coli O157:H7	X	X	✓		▲	✓			✓		✓	▲	▲	▲	
Salmonella	X	X	✓		▲	✓			✓		✓	▲	▲	▲	▲
Shigella			✓			✓	X							✓	X
Yersinia pseudotuberculosis	X							X	✓			✓			
Parasite															
Cyclospora cayetanensis					▲	X	X							✓	X
Virus															
Norovirus					✓	✓	✓				✓			✓	✓
Retorted — **Bacteria**															
Clostridium botulinum			●		●				▲					✓	
Staphylococcus aureus						●									X
Heated — *Bacillus cereus*					X										
BEANS/LEGUMES															
Heated / Refried — **Bacteria**															
Bacillus cereus					X										
Clostridium perfringens			▲		▲	X						▲		▲	

| | | | | | | | | | | | Retail Store/Food Service/Home | | | | | | | | | | |
| Holding/Storage | | | | Processing | | | | | | | Contamination | | | Holding/Storage | | | | Processing | | | |
Improper Hot Holding	Inadequate Refrigeration	Prolonged storage	Room/Outdoor Temperature Holding	Heat Process Failure	Improper Cooling	Improper pH Adjustment	Improper Water Activity (a$_w$)	Inadequate Reheating	Organism/Toxin Survives Process	Improper/Defective Packaging	Cross Contamination	Improper Cleaning of Equipment	Worker/Person	Improper Hot Holding	Inadequate Refrigeration	Prolonged storage	Room/Outdoor Temperature Holding	Heat Process Failure	Improper Cooling	Inadequate Reheating	Organism/Toxin Survives Process
	✓										✓	✓			✓		▲				
											✓	✓			✓	✓	✓				
	✗										▲	✓			✓		▲				
													✗								
													✗								
	✓										✓	▲			✓						
	✓	✗									▲				✓	✓					
	✓										▲	▲			✓		▲				
													✗		▲		▲				
											✓	✓									
													✗								
													✗								
												▲	✗	✓		✗			✓		✓
			✓												✓		✓				▲
		✓													▲	▲	▲				
		✓													▲		▲				
											✓		▲		✓		✓				
	▲												✗		▲						
	✓										✓				✓		✓				
	✓										✓		✗		✓		✓				
		✗	✗																		
											✓		✗								
				✗		✓	▲		✗	✗											✗
	✗	✗					T						▲				▲		▲		
						✓				✗	▲	▲		✗	✗	✓	✗	✗	T	✓	
			✓		✗				✗		▲	▲		✗	✗	✓	✗	✗	T	✓	
			✓		✗				✗		▲	▲		✗	✗	✓	✗	✗	✗		

Key E Situations that contributed to outbreaks of foodborne diseases when fruits, nuts, spices, grains

Legend

✗ = Principal Factor to Consider
✓ = Factor to Consider
▲ = Potential Factor to Consider
● = Source of contamination, but likely to be destroyed during later processing
T = Toxin Survives Heat Processes

Fruits, Nuts, Spices, Grains, or Mushrooms	Farm/Field — Contamination Issues											Processing — Contamination Issues						
	Colonized/Infected/Toxigenic Animals	Animals Access Human Sewage	Animal Feces/Manure	Feed	Sewage	Soil/Grass/Mud	Storage	Water	Worker	Inadequate/Improper Cooling	Storage Conditions	Cross contamination	During Cooling	Environment	Improper Cleaning of Equipment	Manipulation/Spread	Use of Contaminated Water	Worker
Fruits																		
BERRIES — Raw — *Parasite*																		
Cyclospora cayetanensis			▲			✗	✗											✓
Virus																		
Hepatitis A Virus			✗			▲	✗											✗
Norovirus			✓				✗						✓				▲	✗
MELON — Raw — *Bacteria*																		
Escherichia coli O157:H7	✗		✗									▲			▲			
Salmonella	✗		✗		▲	✓	✗	▲				✗			▲	✗	▲	✓
OTHER FRUITS — Raw — *Bacteria*																		
Salmonella	✗		✗		▲	✓	✗	▲				✗			▲	▲	▲	▲
Virus																		
Norovirus					✓		✓	✗						✓			▲	✗
Fruit/Vegetable Juices — Raw — *Bacteria*																		
Escherichia coli O157:H7	✗		✗			▲		▲				✗				✓	✓	
Salmonella	✗		✗			✓		✓				✗				✓	✓	▲
Parasite																		
Cryptosporidium	✗		✗			✗		▲				✗				✓	✓	
Virus																		
Hepatitis A Virus						▲		✓	✗									✗
Processed (Juice HACCP) — *Bacteria*																		
Salmonella	●		●					●	●			✗				▲	▲	▲
Salmonella Typhi			●						●			▲				▲	▲	✗
Virus																		
Hepatitis A Virus			●					●	●								▲	✗
Heat Processed (includes reconstituted) — *Bacteria*																		
Clostridium botulinum			●			●						✗						
Salmonella	●		●					●	●				✓					
Salmonella Typhi			●						●				✓				▲	✗
Virus																		
Hepatitis A Virus			●					●	●									✓

	Holding/Storage				Processing							Contamination				Holding/Storage				Processing			
												Retail Store/Food Service/Home											
	Improper Hot Holding	Inadequate Refrigeration	Prolonged storage	Room/Outdoor Temperature Holding	Heat Process Failure	Improper Cooling	Improper pH Adjustment	Improper Water Activity (a_w)	Inadequate Reheating	Organism/Toxin Survives Process (Processing)	Improper/Defective Packaging	Cross Contamination	During Reconstitution	Improper Cleaning of Equipment	Worker/Person	Improper Hot Holding	Inadequate Refrigeration	Prolonged storage	Room/Outdoor Temperature Holding	Heat Process Failure	Improper Cooling	Inadequate Reheating	Organism/toxin survives
															×								
															×								
		▲		▲								×		▲			✓		✓				
		▲		×								×		▲	▲		✓		×				
		✓		✓								×		▲	▲		✓		✓				
														×									
		✓		▲								×					✓		▲				
		✓		▲								×		▲			✓		▲				
												×											
															×								
		▲		▲						×							▲	▲	▲				
		▲		▲						✓					×		▲	▲	▲				
										✓					×								
		✓		✓	×					×	▲						✓		✓				
		▲		▲	×							▲		▲			▲	▲	▲				
		▲		▲	×							×		×			▲	▲	▲				
					×							▲		✓									

(continued)

Fruits, Nuts, Spices, Grains, or Mushrooms

Legend:
- ✗ = Principal Factor to Consider
- ✓ = Factor to Consider
- ▲ = Potential Factor to Consider
- ● = Source of contamination, but likely to be destroyed during later processing
- T = Toxin Survives Heat Processes

		Farm/Field — Contamination Issues											Processing — Contamination Issues						
		Colonized/Infected/Toxigenic Animals	Animals Access Human Sewage	Animal Feces/Manure	Feed	Sewage	Soil/Grass/Mud	Storage	Water	Worker	Inadequate/Improper Cooling	Storage Conditions	Cross contamination	During Cooling	Environment	Improper Cleaning of Equipment	Manipulation/Spread	Use of Contaminated Water	Worker
NUT/NUT BUTTERS, e.g., Peanuts, Almonds, Pistachios																			
Raw	**Bacteria**																		
	Salmonella	✗		✗			✓		▲	▲			✓		✓	✓	✓		▲
Processed	**Bacteria**																		
	Salmonella	●		●			●		●	●			✗		✗	✗	✓		▲
Retorted	**Bacteria**																		
	Clostridium botulinum						●												
COCONUT																			
Dried	**Bacteria**																		
	Salmonella	✗		✗		▲	✓		▲	▲			✗		✓	▲	▲		▲
SPICES																			
Dried/Fermented	**Bacteria**																		
	Salmonella	✗		✗		▲	✓		▲	▲			✗		✓		✓	▲	▲
GRAINS/RICE (Flour)																			
Raw	**Bacteria**																		
	Escherichia coli STEC/VTEC	✗		✗			✓								▲	✓			
	Salmonella	✗		✗		▲	✓		▲	▲					✓	✓			
	Toxins																		
	Aflatoxin						✗					✗					▲		
Heated	**Bacteria**																		
	Bacillus cereus						✗												
BREADS/PASTRIES/DONUTS																			
Processed	**Bacteria**																		
	Salmonella															✓			✗
	Virus																		
	Norovirus													✓	✓				✗
	Hepatitis A Virus																		✗
PASTA/NOODLE																			
Processed	**Bacteria**																		
	Staphylococcus aureus										●						✓		✗
with eggs	*Salmonella*	●		●			●						▲				▲		▲
MUSHROOMS																			
Raw	**Bacteria**																		
	Salmonella	✗		✗		▲	✗		▲										▲
Retorted	*Clostridium botulinum*						●								▲				
	Staphylococcus aureus									●	●								✗

| | | | | | | | | | | | Retail Store/Food Service/Home | | | | | | | | | | | |
| | Holding/Storage | | | Processing | | | | | | | Contamination | | | | Holding/Storage | | | | Processing | | | |
Improper Hot Holding	Inadequate Refrigeration	Prolonged storage	Room/Outdoor Temperature Holding	Heat Process Failure	Improper Cooling	Improper pH Adjustment	Improper Water Activity (a_w)	Inadequate Reheating	Organism/Toxin Survives Process (Processing)	Improper/Defective Packaging	Cross Contamination	During Reconstitution	Improper Cleaning of Equipment	Worker/Person	Improper Hot Holding	Inadequate Refrigeration	Prolonged storage	Room/Outdoor Temperature Holding	Heat Process Failure	Improper Cooling	Inadequate Reheating	Organism/toxin survives
											✓		▲									
				×							✓		×	▲					×			
				×																		
							▲		×		▲	▲										×
							▲		×		▲	▲										×
											▲											
											✓											
		✓					▲		T													
	×		×				✓		×							×	✓	×		×		×
											✓		×									
											✓		×									
													×									
×	×		×	▲	×				T				✓	×		×		×	▲			T
×	×		×	×	×						▲		▲	▲		×		×	×		×	
														▲								
				×					✓													
			×						T													

Legend:

× = Principal Factor to Consider
✓ = Factor to Consider
▲ = Potential Factor to Consider
● = Source of contamination, but likely to be destroyed during later processing
T = Toxin Survives Heat Processes

Mixed foods	Farm/Field — Contamination					Processing — Contamination Issues						
	Colonized/Infected/Toxigenic Animals	Environment/Climate	Animal Feces/Manure	Soil/Grass/Mud	Worker	Cross contamination	During Cooling	Environment	Improper Cleaning of Equipment	Manipulation/Spread	Use of Contaminated Water	Worker
CHOCOLATE												
Formulated/Blended — **Bacteria**												
Salmonella	●	●	●		●	▲		▲	▲	▲	×	▲
CUSTARD / CREAM-FILLED PASTRY												
Formulated; Formulated/Heated — **Bacteria**												
Salmonella						✓		▲	▲			▲
Staphylococcus aureus						▲			▲			×
FROSTING / TOPPING												
Formulated/Blended — **Virus**												
Hepatitis A Virus												×
Norovirus						▲	✓					×
SOUPS / STEW/ GRAVY												
Formulated/Heated — **Bacteria**												
Bacillus cereus									▲			
Clostridium perfringens									▲			▲
Salmonella						▲			▲			▲
Staphylococcus aureus									▲			×
RICE-BASED DISHES (Fried rice)												
Mixed/Heated — **Bacteria**												
Bacillus cereus				×								
MEAT, VEGETABLE/ CEREAL MIXTURES, e.g., Stir fry, Lasagna, Chinese food, Casseroles												
Mixed/Heated — **Bacteria**												
Clostridium perfringens						▲			▲			▲
Salmonella						▲			▲			▲
Staphylococcus aureus						▲			▲			×
GREEN SALADS WITH PROTEIN, i.e., Vegetable salad with additional processed products, such as												
Mixed/Raw with Portions Heated — **Bacteria**												
Campylobacter						▲						
Escherichia coli STEC\ VTEC						✓		✓	✓	✓	▲	
Salmonella						✓		✓	✓	✓	▲	▲
Shigella						▲					✓	×
Parasite												
Various (such as Cryptosporidium and Giardia)											✓	✓
Virus												
Hepatitis A Virus											✓	×
Norovirus						▲	✓				✓	×

									Retail Store/Food Service/Home										
Holding/Storage				Processing					Contamination			Holding/Storage				Processing			
Improper Hot Holding	Inadequate Refrigeration	Prolonged storage	Room/Outdoor Temperature Holding	Heat Process Failure	Improper Cooling	Improper Water Activity (a_w)	Inadequate Reheating	Organism/Toxin Survives Process	Cross contamination	Improper Cleaning of Equipment	Worker/Person	Improper Hot Holding	Inadequate Refrigeration	Prolonged storage	Room/Outdoor Temperature Holding	Heat Process Failure	Improper Cooling	Inadequate Reheating	Organism/Toxin Survives Process
				×		▲		×	▲	▲	▲								
	×		×	×	×	▲		▲	✓	▲	▲	✓	×		×	×	✓		▲
	×		×	×	×	✓		T	▲	▲	×	✓	×		×	×	✓		T
										×									
									▲	×									
×	×		×		×		×T	×		▲		×	×		×		×T	×	
×	×		×		×		×	×		▲	▲	×	×		×		×	×	
✓	✓		✓	×	✓		✓		▲	▲	▲	✓	✓		✓	×	✓	✓	
×	×		×	▲	▲		✓	T	▲	×	×	×	×		×	▲	✓	✓	T
▲	▲		▲		▲		T✓	▲				×	×		×		×	T✓	×
✓	✓		✓		✓		✓	✓	▲	▲	▲	×	×		×		×	×	✓
✓	✓		✓	✓	✓		✓		▲	▲	▲	×	×		×	×	▲	×	
✓	✓		✓	▲	▲		T	✓	▲	▲	×	×	×		×		✓	T	✓

Cobb salad, Taco salad, Salads with meat, fish, dairy or egg ingredients

	▲		▲	▲	▲				×				▲		▲	▲	✓	▲	
	✓		✓	✓	▲				×	▲			✓		✓	×	✓		
	✓		✓	✓	▲				×	▲	▲		✓		✓	×	✓		
	▲		▲						▲		×		▲		▲				
											✓								
									▲		×								
									▲		×								

(continued)

Mixed foods			Farm/Field					Processing						
✗ = Principal Factor to Consider ✓ = Factor to Consider ▲ = Potential Factor to Consider • = Source of contamination, but likely to be destroyed during later processing T = Toxin Survives Heat Processes					Contamination			Contamination Issues						
			Colonized/Infected/Toxigenic Animals	Environment/Climate	Animal Feces/Manure	Soil/Grass/Mud	Worker	Cross contamination	During Cooling	Environment	Improper Cleaning of Equipment	Manipulation/Spread	Use of Contaminated Water	Worker
MIXED SALADS, i.e., Mixes of proteins, grains and/or vegetables with most ingredients cooked,														
Some Heated Ingredients	**Bacteria**													
		Campylobacter						▲						
		Escherichia coli STEC\ VTEC						✓		✓	✓	✓		
		Salmonella						✗		✓	✓	✓		▲
		Shigella												✗
		Staphylococcus aureus						▲			▲			✗
	Virus													
		Hepatitis A Virus												✗
		Norovirus						▲	✓					✗
VEGETABLE SALADS WITH ONLY RAW INGREDIENTS, e.g., Coleslaw, Garden salad, Caesar														
Mixed / Raw With Portions Heated	**Bacteria**													
		Escherichia coli STEC\ VTEC						✓		✓	✓	✓	▲	
		Salmonella						✓	▲	✓	✓	✓	✓	✓
		Shigella						▲					✓	✗
	Parasite													
		Various (such as Cryptosporidium and Giardia)											✓	✓
	Virus													
		Hepatitis A Virus											✓	✗
		Norovirus						▲	✓				✓	✗
COOKIE DOUGH/CAKE BATTER														
Raw	**Bacteria**													
		Salmonella	✗		✓	✓								
		Escherichia coli O157:H7	✗		✓	✓								

| | | | | | | | | | Retail Store/Food Service/Home | | | | | | | | | | |
| Holding/Storage | | | | Processing | | | | | Contamination | | | Holding/Storage | | | | Processing | | | |
Improper Hot Holding	Inadequate Refrigeration	Prolonged storage	Room/Outdoor Temperature Holding	Heat Process Failure	Improper Cooling	Improper Water Activity (a_w)	Inadequate Reheating	Organism/Toxin Survives Process	Cross contamination	Improper Cleaning of Equipment	Worker/Person	Improper Hot Holding	Inadequate Refrigeration	Prolonged storage	Room/Outdoor Temperature Holding	Heat Process Failure	Improper Cooling	Inadequate Reheating	Organism/Toxin Survives Process

e.g., Chicken salad, Egg salad, Potato salad, Tuna salad, Pasta salad, Salsa with cooked ingredients

	▲		▲	▲	▲				✗				▲		▲	✓	▲		
	✓		✓	✗	▲				✓	✓		▲	✓		✓	✗	▲		
	✓		✓	✗	✓				✗	✓	▲	▲	✗		✗	✗	✓		
	✓		✓	▲	✓						✗	▲	✗		✗				
	✓		✓	▲	✓			✓	▲	▲	✗	▲	✗		✗	▲	✓		✓
											✗								
									▲		✗								

salad, Pico de Gallo, Salsa

	✓		✓						✓	▲			✓		✓				
	✓		✓						✓	✓	✓		✓		✓				
	▲		▲						▲	✗			▲		▲				
											✓								
											✗								
									▲		✗								

Procedures to Investigate Foodborne Illness When Intentional Contamination Is Suspected

Foreword

Historically, intentional contamination of food might not cross the mind of an investigator of foodborne illness. In an era where terrorism is a threat, deliberate contamination needs to be considered whenever a foodborne illness investigation cannot clearly conclude that unintentional contamination was the cause. Much of the investigation of an outbreak or other event involving suspected deliberate contamination of food follows the general procedure for non-intentional contamination of food by pathogens or toxins. But there are some important differences that require specific actions. The format of this updated manual is such that an investigator can use the main body of the manual for common unintentional contamination investigations and this section to prepare for, investigate, and respond to intentional contamination.

Sometimes intentional contamination of food involves a small number of packages of food or limited amounts of exposed food. These instances are often referred to as tampering. Some food tampering has led to foodborne illness. Intentional contamination of food could also involve large quantities of food and/or agents that would result in a large number of illnesses or deaths. This type of scenario may be terrorism-related intentional contamination of food. Regardless of the scale of the event, intentional contamination of food investigations will involve law enforcement, as well as food safety agencies. All samples collected in such investigations will potentially be considered evidence by law enforcement agencies; therefore, chain of custody procedures need to be followed for sample collection and processing.

Develop an Intentional Contamination Surveillance

An effective intentional contamination surveillance system will:

- Be coordinated with existing foodborne disease surveillance systems
- Develop intraagency and interagency emergency response plans
- Be linked to law enforcement and emergency management agencies
- Conduct intraagency and multiagency exercises

Organize the System and Develop Emergency Response Plans

In the face of terrorism events in recent years, local, state or provincial, and federal agencies have prepared, or are in the process of preparing, emergency response plans to address the intentional contamination of foods. Generally, plans are developed within and for an agency with interagency collaboration included. For instance, a public health agency may have its own procedures and protocols for handling intentional food contamination events, but these procedures link with those of other agencies such as law enforcement at the local, state or provincial, and federal levels. Familiarity with the emergency response plans of the jurisdiction, your role and responsibilities, and where you fit within the chain of command for your agency in an intentional contamination event is critical. Your agency may have one emergency plan or procedure for unintentional contamination and a different plan for dealing with intentional contamination and or terrorism events. Unlike unintentional contamination where just food safety/public health agency emergency plans are implemented, multiple agencies may implement their emergency plans when intentional contamination is suspected or confirmed. Coordination among the agencies is critical.

- Develop appropriate emergency response plans to protect public health when preparing for an intentional contamination incident. The plans at a minimum should include sections on introduction, including mission, purpose and scope, planning assumptions, responsibilities and organizations including relationships to law enforcement and emergency management, operating procedures including information flow, risk assessment/situation analysis, activating resources, deactivation, and recovery.
- Emergency response plans should include interagency coordination and information sharing. The plans should specifically detail who contacts other agencies and should include names, phone numbers, after hours contact numbers, e-mail addresses, fax numbers, cellular phone numbers, satellite phone numbers, and Blackberry pin numbers. Keep the contact lists current by updating every 6 months.
- Response plans need to include the roles and responsibilities of the units within the organization and clearly state who is coordinating the emergency response. Conduct an exercise within your organization and then conduct exercises with the agencies you would work with in the event of a real emergency.
- Review your legal authorities and procedures for embargo, detention, and seizure for appropriateness in dealing with intentional contamination. Review and/or develop sampling protocols with a chain of custody procedure for intentional contamination.

Establish an Intentional Contamination Investigation Team

- Develop working relationships with agencies you may be working with in an emergency response mode. Intentional contamination should trigger multiple emergency response plans at all levels of government including local city/county, state or provincial, and the federal or national level. Public health, agriculture, law enforcement, environmental, and emergency management agencies all may become involved in intentional contamination investigations.

• Meet with the law enforcement community; local, state/provincial, federal/national, you will work with on incidents of intentional contamination. Describe the investigatory, public health, and epidemiological role you will play. The law enforcement agency will be the lead in any criminal investigation. Try to establish periodic meetings that you each may host. This helps build confidence with each other prior to an emergency.

• Develop working relationships with the local/state/provincial/federal level emergency management responders and directors. Become familiar with their roles and responsibilities and make them aware of yours. Determine whether they will provide your agency with a workstation at the Emergency Operations Center (EOC) during emergencies involving intentional contamination.

• Inform the emergency response and law enforcement communities of the laboratory capabilities of your agency. If you do not have laboratory capability, for the safety of your employees, do not offer services until your staff has received adequate training in hazardous material collection and sampling. Identify alternative laboratories for use when your laboratory is not able to perform the required tests.

Develop contingency plans on whether you plan to sample and/or analyze the food in your laboratory. Determine whether you will be a first responder in collecting food samples. Train your staff in handling hazardous materials if you know you are handling a deliberately contaminated product as the contaminants involved in intentional contamination may pose some unique hazards. Develop a chain of custody procedure for collecting food samples and documenting the method of collection, sample preparation, storage, and control. Follow these procedures during the incident. In a potential criminal case, following a chain of custody procedure is essential in supporting the law enforcement component of the investigation.

Develop information sharing protocols with law enforcement, emergency management, and other agencies. These protocols should include a level of sensitivity as to whom this information can be shared with in an emergency. For example, who can see confidential, law enforcement sensitive, or classified information and what procedures must you follow to receive clearance to view this type of information?

Determine if your local, state/provincial or national government has a Food Emergency Response Team relevant to this situation, and if they do, have your roles and responsibilities been outlined? Meet with members of the team to review roles and responsibilities and to conduct exercises. These exercises provide experience in following chain of command, implementing procedures and will allow identification of gaps or lapses that need to be corrected prior to an actual event.

Obtain reference materials relevant to different agents and to intentional contamination. Seek out this information from your national and state/provincial food authorities. Another source of this information is your federal or national health authority particularly as it relates to infectious disease agents. The Internet can be used to search for information regarding toxic and infectious agents.

Meet with appropriate industry officials and express concern that the industry should be making plans to respond to a deliberate contamination incident. Offer to listen to their needs and become familiar with their concerns.

Triggers for Recognizing Intentional Contamination of Food

There are a number of challenges in determining whether particular clues are a tip off to an intentional food contamination event. These events may involve diseases that are often investigated, may mimic characteristics of diseases often investigated, or may be hard to recognize if rare, nonendemic, representative of an eradicated disease or weaponized organism is used. Some practitioners may be reluctant to report symptoms until a true diagnosis has been made. Making the determination of an intentional event carries with it concerns of overreacting, creating additional burden within the workplace, or further diminishing scarce resources if the event proves to be a hoax or is later determined to be a routine investigation.

Triggers that can guide the determination of an intentional food contamination event generally fit in one of two categories, public health and law enforcement. Many public health triggers may reflect aspects of an unintentional event. Determination of an intentional food contamination event will likely involve a combination of public health factors and will require consultation with epidemiologists, laboratorians, physicians, and other health professionals.

Questioning whether any one factor or combination of factors that rise to the level of triggering suspicion that an event may be intentional should be encouraged. It is important to understand the types of factors that warrant raising this question in consultation with other public health professionals. Factors that warrant consultation with others include but are not limited to:

- Reports of unusual color, odor, or appearance of food
- Evidence of tampering in food packaging
- Unusual agent or vehicle
- Multiple unusual or unexplained disease entities in one patient
- High attack rate, severe outcome or deaths
- Failure of patients to respond to traditional treatment
- Multiple exposure sites or vehicles with no apparent common link
- Many ill persons presenting near the same time
- Deaths or illness among animals that may be unexplained and precede reports of illness among the human population

Law enforcement officials may also provide information to assist public health officials in deciding how to approach a food complaint or illness. Law enforcement may have intelligence or threat information such as the unlawful possession of agents by any individual or group, indications of a credible threat in an area, or identification and/or seizure of literature pertaining to the development or dissemination of agents. Receipt of information from law enforcement officials should be shared with other public health officials according to the emergency response plan of the jurisdiction and the rules governing the sharing of the information.

Obtain Specific Assistance

In instances of possible intentional contamination, the agencies involved will broaden beyond those customarily involved in unintentional contamination investigations. Law enforcement, emergency medical, agriculture hazardous materials teams, environmental, fire, and homeland security agencies may become involved as well. Once intentional contamination is suspected, law enforcement agencies become the lead investigation agency as criminal activity is involved. As soon as you suspect intentional contamination, notify the persons in charge of your agency and discuss notification of other appropriate agencies. This may include law enforcement, as well as state/provincial and federal/national agencies involved in food safety. The investigation will become a joint one involving public health and law enforcement agencies working together.

Find and Interview Additional Cases

This stage of the investigation bears similarities to steps taken during the investigation of unintentional events. Once hypotheses generation concludes that an intentional event is likely, the investigation takes on a law enforcement context. In addition to contacting health agencies, hospital emergency rooms, and local physicians to find additional cases, contact poison control centers, schools, and major employers in the area that may also provide information on additional cases.

Reinterview of cases may be required. Forms C1 and C2 may be used but will likely require some modification. Specific circumstances of the event will determine the type of questions that may be added. For instance, cases might be asked if they overheard any unusual comments by servers or others, or if they noticed any unusual events or practices.

The number of persons interviewed will depend on circumstances of the event and resources available. In the case of unintentional contamination, questionnaires may be sent to cases and persons at risk for them to answer and return. However, in intentional events, it may become imperative for information to be collected by phone or by face-to-face interview. Public health and law enforcement investigators may conduct these interviews jointly. After questionnaires have been completed, summarize the data. Forms D1 and D2 can be used as models but will likely require modification.

Seek Sources and Modes of Intentional Contamination and Ways by Which the Contaminants Survived and/or Proliferated

Investigation of potential intentional contamination of food calls for different thinking than that used in unintentional contamination investigations. Investigators need to think about how perpetrators would have access to food and how an agent may be introduced into the vehicle. Ways that the agent was mixed or distributed in the vehicle and how it might also have been spread in the environment where the vehicle was located should be considered. Ask experts about the amount of agent, i.e., toxic or infectious dose that would have to been introduced. Obtain samples of any remaining vehicle, ingredients, and the environment where the food was prepared and stored.

Changes in color, odor, or appearance of the vehicle may be indications that a contaminant was deliberately added. Other unusual findings during a food preparation review may identify a chemical that does not belong there, sick or dead animals in the vicinity of the food preparation facility, or unusual equipment or devices such as test tubes or lab equipment in the food preparation facility. Ask if there is new staff or disgruntled employees or contentious labor negotiations. Identify any recent changes in routine procedures, e.g., new suppliers, delivery services, etc. Ask if background checks have been done on the staff and if so what the results were. Ask if the firm has conducted a vulnerability assessment of the potential for intentional contamination and, if so, ask to review it for possible ways that the vehicle could have been contaminated.

Make Recommendations for Control

Seize, detain (embargo), stop distribution, remove, recall, reject, or destroy the epidemiologically implicated lot.

Public health authorities are responsible for taking appropriate control actions at the facility where the epidemiologically implicated food was prepared, eaten, or purchased and other places a traceback investigation led. Implicated product may be seized, detained (embargo), stop distribution, removed, recalled, rejected, or destroyed all of which are considered routine control actions. Handling or disposing of foods that have been seized or detained during the course of an investigation that is eventually recognized to be an intentional contamination event will require consultation with various experts assembled on the emergency response team. Special procedures may be required to handle or destroy these intentionally contaminated foods. If foods have been destroyed prior to recognizing that they were associated with an intentional contamination event, a review of the process with the emergency response team or other experts should be conducted.

Cease processing or preparation of the epidemiologically implicated food.

The facility where the food was prepared, stored, eaten, or purchased may require special decontamination or disposal procedures depending on the agent involved in an intentional food contamination event. Consultation with the emergency response team or other experts will be required. The decontamination/disposal activities may or may not extend beyond those considered within the scope of routine public health practice. While the public health agency may not be responsible for overseeing a decontamination/disposal activity in a facility, officials should know under what circumstances the facility can resume its normal activities and what documentation of the decontamination/disposal should be required. In many jurisdictions, the agency responsible for environmental protection will have a lead role in decontamination and disposal. The environmental protection agency and the public health agency will need to work together to assure protection of public health and protection of the environment.

Determine your weakest areas in your food operation in order to protect the integrity of your food supply by using CARVER + Shock.

Conduct a CARVER + Shock assessment of your food production facility or process, to identify resources on protecting the most susceptible points in your system.

Food industries can be proactive in reducing the risk of deliberate contamination of their food supplies by using this software tool developed by the U.S. Food and Drug Administration. CARVER is an acronym for the following six attributes used to evaluate the attractiveness of a target for attack:

- Criticality – measure of public health and economic impacts of an attack
- Accessibility – ability to physically access and egress from target
- Recuperability – ability of the system to recover from an attack
- Vulnerability – ease of accomplishing an attack
- Effect – amount of direct loss from an attack as measured by loss in production
- Recognizability – ease of identifying target

A seventh attribute, Shock, has been added to the original six to assess the combined health, economic and psychological impacts of an attack within the food industry. This tool can be used to assess the vulnerabilities within a system or infrastructure in the food industry. The software mimics the thought processes in play during a face-to-face CARVER + Shock session by having the user to: 1) build a process flow diagram for the system to be evaluated, and 2) answer a series of questions for each of the seven CARVER + Shock attributes for each process flow diagram node. Each question has an associated score. Based on the answers given, the software calculates a score for each CARVER + Shock attribute and sums them to produce a total score for each node. In this way this independent evaluation may point out weaknesses in the company infrastructure and suggest ways of closing gaps where a disgruntled employee or terrorist could contaminate food or its ingredients.

Specific Advice in Informing the Public about a Suspected or Confirmed Intentional Contamination Event

Multiagency teams will be involved in investigating an intentional contamination event. The food protection authorities will need to review and participate in developing media releases to the public. The food protection authority may not be the lead in this event, so it is critical to develop working relationships with the agencies most likely involved in advance of any event. In a multiagency event, the group should identify one spokesperson for all agencies to address the media. The lead spokesperson can serve to ensure the accuracy of the information being released while safeguarding the release of highly sensitive information. The lead spokesperson must work with representatives of all participating agencies to assure that appropriate information that is cleared by all agencies is provided to the public.

Further Readings

Allos, B.M., M.R. Moore, P.M. Griffin, and R.V. Tauxe. 2004. Surveillance for sporadic foodborne dsease in the 21st century: The FoodNet perspective. Clin. Infect. Dis. 38(3):S115-S120. http://cid.oxford-journals.org/search?author1=Ban+Mishu+Allos&sortspec=date&submit=Submit http://cid.oxford-journals.org/search?author1=Matthew+R.+Moore&sortspec=date&submit=Submit http://cid.oxfordjournals.org/search?author1=Patricia+M.+Griffin&sortspec=date&submit=Submit http://cid.

oxfordjournals.org/search?author1=Robert+V.+Tauxe&sortspec=date&submit=Submit   http://cid.
oxfordjournals.org/content/38/Supplement_3.toc

Batz, M.B., M.P. Doyle, J.G. Morris, J. Painter, R. Singh, R.B. Tauxe, M.R. Taylor, and D.M.A. Lo Fo
Wong. 2005. Attributing illness to food. Emerg. Infect. Dis. 111:993–999. (see link to journal below)

Bloomfield, S.F., and E.A. Scott. 2003. Developing an effective policy for home hygiene: a risk-
based approach. Intl. J. Environmental Health. June: 13 Suppl 1:S57–66.

Boore, A., K.M. Herman, A.S. Perez, C.C. Chen, D.J. Cole, B.E. Mahon, P.M. Griffin, I.T.
Williams, and A.J. Hall. 2010. Surveillance for foodborne disease outbreaks — United States,
2007. Morbidity & Mortality Weekly Report. 59(31):973-979.

Bryan, F.L. 2002. Reflections on a career in public health: evolving foodborne pathogens, environmental health, and food safety programs. J. Environ. Health. 2002 Dec;65(5):14–24.

Bryan, F.L., E.C.D. Todd, and J.J. Guzewich. 1997. Surveillance of foodborne diseases II.
Summary and presentation of descriptive data and epidemiologic patterns; their value and
limitations. J. Food Prot. 60:567–578.

Bryan, F.L., J.J. Guzewich, and E.C.D. Todd. 1997. Surveillance of foodborne diseases III.
Summary and presentation of data on vehicles and contributory factors; their value and limitations. J. Food Prot. 60:701–714.

(CIFOR) Council to Improve Foodborne Outbreak Response, 2009. Foodborne Disease Surveillance
and Outbreak Detection, Guidelines for Foodborne Disease Outbreak Response, USA. Available
at: http://www.cifor.us/CIFORGuidelinesProjectMore.cfm (Accessed March 29, 2011).

Chittick, P., A. Sulka, R.V. Tauxe, and A.M. Fry. 2006. A summary of national reports of foodborne outbreaks of *Salmonella* Heidelberg infections in the United States: clues for disease
prevention. J. Food Prot. 69(5):1150–1153.

Dalton, C.B., T.D. Merritt, L.E. Unicomb, M.D. Kirk, R.J. Stafford, K. Lalor, and the OzFoodNet
Working Group. 2010. A national case-control study of risk factors for listeriosis in Australia.
Epidemiol. Infect. 139(3):437–445.

Doyle, M.P., G.R. Acuff, D. Bernard, B. Cords, J. Cullor, J. Hollingsworth, K.P. Penner, S.
Seward, W.A. Sperber, and B. Tompkin. 2004. Intervention strategies for the safety of foods
of animal origin. Council for Agricultural Science and Technology (CAST). Issue paper No.
25 (see link below under website).

Doyle, M.P., and L.R. Beuchat (eds.). 2007. *Food microbiology: fundamentals and frontiers*. (3rd
edition). ASM Press, Washington, D. C.

Farber, J.M. and E.C.D. Todd (eds.). 2002. *Safe Food Handling*. Marcel Dekker: New York. ISBN
0824703316.

Food and Drug Administration. 2011. CARVER + Shock. New release. Available at: http://www.
fda.gov/Food/FoodDefense/CARVER/default.htm. Accessed on April 27, 2011.

Frenzen, P.D., A. Drake, and F.J. Angulo: Emerging Infections Program FoodNet Working Group.
2005. Economic cost of illness due to *Escherichia coli* O157 infections in the United States.
J. Food Prot. 68(12):2623–2630.

Greig, J.D., E.C.D. Todd, C.A. Bartleson, and B.S. Michaels. 2007. Outbreaks where food workers have been implicated in the spread of foodborne disease. Part 1. Description of the problem, methods, and agents involved. J. Food Prot. 70:1752–1761.

Guzewich, J.J., F.L. Bryan, and E.C.D. Todd. 1997. Surveillance of foodborne diseases I. Purpose
and types of surveillance systems and networks. J. Food Prot. 60:555–566.

Headrick, M.L., S. Korangy, et al. 1998. The epidemiology of raw milk-associated foodborne
disease outbreaks reported in the United States, 1973 through 1992. American Journal of
Public Health 88(8): 1219–1221.

Heymann, David L. (ed.) 2004. *Control of Communicable Diseases Manual* (18th edition).
American Public Health Association. ISBN: 0-87553-034-6

Hrudey, S.E., and E.J. Hrudey. 2004. *Safe Drinking Water.* IWA Publishing (London) ISBN 1843390426

ILSI Research Foundation Risk Science Institute Expert Panel on *Listeria monocytogenes* in
foods. 2005. Achieving continuous improvement in reductions in foodborne listeriosis –
A risk-based approach. J. Food Prot. 68:1932–1994.

Jay, J.J., M.J. Loessner, and D.A. Golden. 2006. *Modern Food Microbiology* (7th Edition).
Springer. ISBN: 0387231803.

Jaykus, L., and S. Dennis (Co-Chairs). Using Risk Analysis to Inform Microbial Food Safety Decisions. 2006. Council for Agricultural Science and Technology (CAST). Issue Paper No. 31. (see link below under Websites)

Kirk, M.D., M.G. Veitch and G.V. Hall. 2010. Gastroenteritis and food-borne disease in elderly people living in long-term care. Clin. Infect. Dis. 50(3):397–404.

Kuchenmüller, T., S. Hird, C. Stein, P. Kramarz, A. Nanda, and A.H. Havelaar. 2009. Perspectives: Estimating the global burden of foodborne diseases - a collaborative effort. Eurosurveillance 14(18):1–4.

Majowicz, S.E., J. Musto, E. Scallan, F.J. Angulo, M. Kirk, S.J. O'Brien, T.F. Jones, A. Fazil, and R.M. Hoekstra. 2010. The global burden of nontyphoidal Salmonella gastroenteritis. Clin. Infect. Dis. 50(6): 882–889.

Mead, P.S., L. Slutsker, V. Dietz, L.F. McCaig, J.S. Bresee, C. Shapiro, P.M. Griffin, and R.v. Tauxe, 1999 Food-Related Illness and Death in the United States. Emerg. Infec. Diseases. 5(5):607–625.

Meftahuddin, T. 2002. Review of the trends and causes of food borne out-breaks in Malaysia from 1988 to 1997. Med. J. Malaysia 57(1):1–2.

Panico, M.G., V. Caporale, and E. Agrozzino. 2006. Investigating of a foodborne outbreak: analysis of the critical points. Ann. Ig. 2006 May–Jun; 18(3):191–197.

Riemann, H.P., and D.O. Cliver, (eds.) 2006. Foodborne Infections and Intoxications (3rd Edition). Elsevier: Amsterdam. ISBN 13 9780125883658.

Ryser, E.T., and E. Marth, (eds.) 2007, Listeria, Listeriosis and Food Safety. (3rd Edition). Taylor and Francis, Boca Raton, FL.

Scallan E., R.M. Hoekstra, F.J. Angulo et al. 2011. Foodborne illness acquired in the United States — major pathogens. Emerg. Infect. Dis. 17:7–15.

Scallan E., P.M. Griffin, F.J. Angulo, R.V. Tauxe, R.M. Hoekstra. 2011. Foodborne illness acquired in the United States — unspecified agents. Emerg. Infect. Dis. 17:16–22.

Swaminathan, B., T.J. Barrett, and P. Fields. 2006. Surveillance for human *Salmonella* infections in the United States. J. A.O.A.C. Int. 89(2):553–559.

Thompson, S., R. de Burger, and O. Kadri. 2005. The Toronto food inspection and disclosure system: A case study. British Food Journal 7(3):140–149.

Todd, E.C.D., J.J. Guzewich, and F.L. Bryan. 1997. Surveillance of foodborne disease IV. Dissemination and uses of surveillance data. J. Food Prot. 60:715–723.

Todd, E.C.D. 2006. Challenges to global surveillance of disease patterns. Marine Pollution Bulletin 53:569–578.

Todd, E.C.D., J.D. Greig, C.A. Bartleson, and B.S. Michaels. 2007. Outbreaks where food workers have been implicated in the spread of foodborne disease. Part 3. Factors contributing to outbreaks and description of outbreak categories. J. Food Prot. 70:2199–2217.

Todd, E.C.D., J.D. Greig, B.S. Michaels, C.A. Bartleson, D. Smith, and J. Holah. 2010. Outbreaks where food workers have been implicated in the spread of foodborne disease. Part 11. Use of antiseptics and sanitizers in community settings and issues of hand hygiene compliance in health care and food industries. J. Food Prot. 73:2306–2320.

Vaillant, V., H. De Valk, E. Baron et al. 2005. Foodborne infections in France. Foodborne Pathog. Dis.2(3):221–232.

Wong, L.F., J.K. Andersen, B. Nørrung, H.C. Wegener. Food contamination monitoring and foodborne disease surveillance at national level. Building Effective Food Safety Systems, Second FAO/WHO Global Forum of food Safety Regulators, Bangkok, Thailand, 12-14 October, 2004, Food and Agriculture Organization, Rome, Italy.

WHO (World Health Organization), Surveillance Programme for Control of Foodborne Infections and Intoxications in Europe. 8th Report. 1999–2000. 2004. http://www.bfr.bund.de/internet/8threport/8threp-fr.htm.

Year in review: Communicable Disease Surveillance, 2005. New South Wales (NSW) Public Health Bull. 17(5–6):65–75. http://www.health.nsw.gov.au/public-health/phb/HTML2006/mayjun06html/article1p65.html.

WHO. Foodborne Disease Outbreaks. Guidelines for Investigation and control 2007. Geneva, Switzerland. http://www.who.int/foodsafety/publications/foodborne_disease/fdbmanual/en/index.html (accessed 12/24/2007).

Table A Equipment useful for investigations[a]

Item	Examples
Investigation guidelines and investigative forms	IAFP manual, "Procedures to Investigate Foodborne Illness, 6th ed"; 50 copies of Form C; one dozen copies each of Forms E and F; six copies of Forms G, H, I, J; three copies of Forms D, M, and J; and one copy each of Forms L, N, O, and P. Epi-Info™ software.
Sterile sample containers	Plastic bags (disposable or Whirl-Pak® type, or stomacher), wide-mouth jars (6 oz to 1 qt capacity) with screw caps, water sample bottles (bottles for chlorinated water should contain enough sodium thiosulfate to provide a concentration of 100 mg of this compound per ml of sample), foil or heavy wrapping paper, metal cans with screw-type caps.
Sterile and wrapped sampling implements	Spoons, tongs, scoop, tongue-depressor blades, butcher knife, forceps, spatula, drill bits, metal tubes 15–30 cm (½–1 in.) in diameter, 360–720 cm (1–2 ft) long, pipets, scissors, swabs, sponges. Moore swabs (compact pads of gauze, made from 1,440 cm [4 ft] × 180 cm [6 in.] strip, tied in center with a long, stout twine or wire – for sewer, drain, stream or pipeline sampling).
Specimen-collecting equipment (for human specimens from cases and controls)	Containers (with lids) for stool specimens, bottles containing a bacterial preservative and transport medium, mailer tubes or styrofoam box, sterile swabs, rectal swab units, tubes of bacterial transport medium, stool preservative medium for parasites, phlebotomy supplies for blood specimens.
Disinfection and sterilizing agents	0.5% w/v solution of calcium hypochlorite or 5.25% household liquid bleach; 95% ethyl alcohol, propane torch.
Refrigerants	Canned ice, refrigerant in plastic bags, liquid in cans, rubber, plastic bags which can be filled with water and frozen, heavy-duty plastic bags for ice.
Media	Transport media, preenrichment or enrichment broth, as appropriate.
Hazard analyses/ Temperature, pH and a_w measuring	Hand-held potentiometer, data logger or strip chart recorder; thermocouples with needle-end sensors of varying lengths, welded-end sensors, and surface sensors; radiation "gun" potentiometer, bayonet-type thermometers 0 to 220°F (−20 to 110°C) 5 in. (13 cm) long, and 8 in. (20 cm) long, in protective case; watch; pH meter with pH 7 and 4.01 buffers; a_w meter; disinfectant test kit(s) and/or papers; graph paper; plastic drawing angles and curves. All equipment should be calibrated.
Clothing (optional)	Coveralls, apron or laboratory coat; either paper hats, hard hats or hair nets; disposable plastic gloves.
Supporting equipment	Lap-top or notebook computer, computer software, blank disks; sterile plastic gloves; plastic container liners for ice; water-proof marking pens; pencils; note pad; roll of adhesive or masking tape; labels; water-proof card-board tags with eyelets and wire ties; flashlight; matches; test tube rack to fit tubes used; insulated chest or styrofoam container; packing material; camera with flash attachment; spare batteries for all equipment in which they are used.
Statistical calculations	Computer with software programs to do chi square, Fisher's exact, Odds Ratio and Relative Risk probability, and other calculations.

[a]Assemble a kit to be kept at any agency responsible for investigating foodborne illness. It should include: at least 15 sterile plastic bags or wide mouth jars, 15 spoons, 6 specimen collection containers or devices, and one each of the supporting equipment and sterilizing or disinfecting equipment should be preassembled. Periodic resterilization or replacement of sterile supplies, media, or transport media is required to maintain the kit in a ready-to-use condition. "Sterile" tape should be attached to all sterilized objects or they should be marked with the date of sterilization. Microbiological kits have expiration dates and need to be replaced before that date; should not be used after kit expiration date has been reached

Table B Illnesses acquired by ingestion of contaminated foods: A condensed classification by symptoms, incubation periods, and types of agents

Illness	Etiologic agent and source	Incubation or latency period[a]	Signs and symptoms[a]	Foods usually involved[b]	Specimen to collect	Factors contributing to foodborne outbreaks[d]
			UPPER GASTROINTESTINAL SIGNS AND SYMPTOMS [NAUSEA, VOMITING] PREDOMINATE			
			Incubation (latency) period usually less than 1 h			
			Fungi			
Gastrointestinal irritating group mushroom poisoning	Possibly resin-like substances in some mushrooms (mushroom species are different from those cited on pages 103 and 113)	30 min to 2 h	Nausea, vomiting, retching, diarrhea, abdominal pain	Many varieties of wild mushrooms	Vomitus	Eating unknown varieties of wild mushrooms; mistaking toxic mushrooms for edible varieties
			Chemicals			
Antimony poisoning	Antimony in gray enamelware	Few minutes to 1 h	Vomiting, abdominal pain, diarrhea	High-acid foods and beverages	Vomitus, stools, urine	Purchasing/using antimony-containing utensils; storing high-acid foods in chipped gray enamelware
Cadmium poisoning	Cadmium in plated utensils	15–30 min	Nausea, vomiting, abdominal cramps, diarrhea, shock	High-acid foods and beverages; metal-colored cake decorations	Vomitus, stools, urine, blood	Purchasing/using cadmium-containing utensils; storing high-acid beverages in cadmium containers
Copper poisoning	Copper in pipes and utensils; old ice cream machines; old dairy white metal	Few minutes to few hours	Metallic taste, nausea, vomiting (green vomitus), dizziness, abdominal pain, diarrhea, chills	High-acid foods and ice creams (ices) and beverages	vomitus, gastric washing, urine, blood	Faulty backflow preventors in vending machines or soda fountains; storing or vending high-acid (low pH) beverages from copper containers, pipe lines, or old equipment containing copper
Fluoride poisoning	Sodium fluoride in insecticides and rodenticides	Few minutes to 2 h	Salty or soapy taste, numbness of mouth, vomiting, diarrhea, dilated pupils, spasms, pallor, shock, collapse	Any accidentally-contaminated foods, particularly dry foods (such as dry milk, flour, baking powder, cake mixes)	Vomitus, gastric washing	Storing insecticides in same area as foods, mistaking pesticides for powdered foods

(continued)

Table B (continued)

Illness	Etiologic agent and source	Incubation or latency period[a]	Signs and symptoms[a]	Foods usually involved[b]	Specimen to collect	Factors contributing to foodborne outbreaks[d]
Lead poisoning	Lead in earthen-ware vessels; pesticides, paint, plaster, putty, soldered joints	30 min or longer	Metallic taste, burning of mouth, abdominal pain, milky vomitus, bloody or black stools, foul breath, blue gum line, shock	High-acid foods and beverages stored in lead-containing vessels; any accidentally contaminated food	Vomitus, gastric washing, stools, blood, urine	Purchasing or using lead-containing vessels; storing high-acid foods including wine in lead-containing vessels; storing pesticides in same area as food
Tin poisoning	Tin in tinned cans or containers	30 min to 2 h	Bloating, nausea, vomiting, abdominal cramps, diarrhea, headache	High-acid foods and beverages	Vomitus, gastric washing, urine, blood, stools	Storing high-acid foods in tinned cans or containers in which there is no lacquer or the lacquer had peeled. Very high concentrations are required to cause illness
Zinc poisoning	Zinc in galvanized containers	Few minutes to few hours	Pain in mouth and abdomen, nausea, vomiting, dizziness	High-acid foods and beverages	Vomitus, gastric washing, urine, blood, stools	Storing high-acid foods in galvanized cans
Bacteria						
Incubation (latency) period usually between 1 and 6 h						
Bacillus cereus gastroenteritis (emetic type)	Exo-enterotoxin of *B. cereus*; organism in soil (strains differ from those cited on page 105)	½–5 h	Nausea, vomiting, occasionally diarrhea	Boiled or fried rice, cooked corn-meal dishes, porridge, pasta	Vomitus, stools	Storing cooked foods at room temperature; storing cooked foods in large containers in refrigerator; preparing foods several hours before serving
Staphylococcal intoxication	Exoenterotoxins A, B, C, D, E, F or H of *Staphylococcus aureus*. Staphylococci from nose, skin and lesions of human beings and other animals and from udders of cows	1–8 h, typically 2–4 h	Nausea, vomiting, retching, abdominal pain, diarrhea, prostration	Ham, meat or poultry products; cream-filled pastries; whipped butter; cheese; dry milk; high protein leftover foods	Ill: vomitus, stools, rectal swabs. Food handlers: nasal swabs, swabs of lesions	Storing cooked foods at room temperature; storing cooked foods in large containers in refrigerator; touching cooked foods; preparing foods several hours before serving; holding foods at warm bacterial-incubation temperatures; fermentation of abnormally low-acid foods; handling foods by persons with pus-containing infections

Chemicals

Name	Causative agent	Incubation	Symptoms	Foods	Specimen	How contaminated
Nitrite poisoning[c]	Nitrites or nitrates used as meat curing compounds	1–2 h	Nausea, vomiting, cyanosis, headache, dizziness, weakness; loss of consciousness; chocolate-brown colored blood[c]	Cured meats; any accidentally-contaminated food; spinach exposed to excessive nitrification	Blood	Using excessive amounts of nitrites or nitrates in foods for curing or for covering up spoilage; mistaking nitrites for common salt and other condiments; improper refrigeration of fresh produce; excessive nitrification of fertilized foods

Marine Phytoplankton

Name	Causative agent	Incubation	Symptoms	Foods	Specimen	How contaminated
Diarrhetic shellfish poisoning	Okadaic acid and other toxins produced by dinoflagellates *Dinophysis* spp.	½–12 h, usually 4 h	Diarrhea, nausea, vomiting, abdominal cramps, chills	Mussels, clams, scallops	Gastric washing	Harvesting shellfish from waters with higher than usual concentration of *Dinophysis* spp.
Azaspiracid shellfish poisoning (AZP) (a few outbreaks from Irish mussels to date)	Azaspiracids (polyethers) from dinoflagellate *Azadinium*, and it predator *Protoperidinium* (West European waters to date).	Minutes to hours	Nausea, vomiting, severe diarrhea, and stomach cramps were similar of those of diarrhetic shellfish poisoning (DSP); may target multiple organs and be a carcinogen	Mussels, scallops, clams, crabs	Gastric washings or vomitus to detect the dino-flagellate species	Eating shellfish from toxic areas where *Azadinium* dinoflagellates occur

Incubation (latency) period usually between 7 and 12 h

Fungi

Name	Causative agent	Incubation	Symptoms	Foods	Specimen	How contaminated
Cyclopeptide and gyromitrin groups of mushroom poisoning	Cyclopeptides and gyromitrin in some mushrooms (mushroom species are different from those cited on pages 100 and 113)	6–12 h	Abdominal pain, feeling of fullness, vomiting, protracted diarrhea, loss of strength, thirst, muscle cramps, collapse, jaundice, drowsiness, dilated pupils, coma; death	*Amanita phalloides, A. verna, A. bisporigera, Galerina autumnalis, Gyromitra esculenta* (false morels) and similar species of mushrooms	Urine, blood, vomitus	Eating certain species of *Amanita, Galerina,* and *Gyromitra* mushrooms; eating unknown varieties of mushrooms; mistaking toxic mushrooms for edible varieties

(continued)

Table B (continued)

Illness	Etiologic agent and source	Incubation or latency period[a]	Signs and symptoms[a]	Foods usually involved[b]	Specimen to collect	Factors contributing to foodborne outbreaks[d]
			Incubation (latency) period between 13 and 72 h			
			Viruses			
Norovirus infection or gastroenteritis	Caliciviruses	Typically 24–48 h	Nausea, vomiting, diarrhea, abdominal pain, myalgia, headache, malaise, low-grade fever; duration up to 60 h	Human feces	Stools, vomitus	Infected persons touching ready-to-eat foods; harvesting shellfish from sewage polluted waters; inadequate sewage disposal; using contaminated water
			BURNING MOUTH, SORE THROAT AND/OR RESPIRATORY SYMPTOMS AND SIGNS OCCUR			
			Incubation period less than 1 h			
			Chemicals			
Calcium chloride poisoning	Calcium chloride freezing mixture for frozen dessert bars	Few minutes	Burning lips, mouth, throat; vomiting	Frozen dessert bars	Vomitus	Splashing of freezing mixture onto popsicles while freezing; cracks in molds allowing $CaCl_2$ to penetrate popsicle syrup
Sodium hydroxide poisoning	Sodium hydroxide in bottle-washing compounds, detergents, drain cleaners, or hair straighteners	Few minutes	Burning of lips, mouth and throat; vomiting, abdominal pain, diarrhea	Bottled beverages, pretzels	Vomitus	Inadequate rinsing of bottles cleaned with caustic soda; inadequate baking of pretzels
			Incubation (latency) period usually between 18 and 72 h			
			Bacteria			
Beta-hemolytic streptococcal infections	*Streptococcus pyogenes* from throat and lesions of infected humans	1–3 days	Sore throat, fever, nausea, vomiting, rhinorrhea; sometimes a rash. Sequela: rheumatic fever	Raw milk, egg-containing salads	Throat swabs, vomitus	Persons touching cooked foods; touching of foods by persons with pus-containing infections; room-temperature storage; storing cooked foods in large containers in refrigerator; inadequate cooking or reheating; preparing foods several hours before serving

LOWER GASTROINTESTINAL TRACT SIGNS AND SYMPTOMS [ABDOMINAL CRAMPS, DIARRHEA] PREDOMINATE

Incubation (latency) period usually between 7 and 17 h

Bacteria

Bacillus cereus enteritis (diarrheal type)	Enterotoxins of *B. cereus*. Organisms in soil (strains differ from those cited on page 102)	8–16 h, mean 12 h	Nausea, abdominal pain, watery diarrhea	Cereal products, soups, custards and sauces, meatloaf, sausage, cooked vegetables, reconstituted dried potatoes, refried beans	Stools	Storing cooked foods at room temperature; storing cooked foods in large containers in refrigerator; holding foods at warm (bacterial-incubating) temperatures; preparing foods several hours before serving; inadequate reheating of leftovers
Clostridium perfringens enteritis	Endoenterotoxin formed during sporulation of *C. perfringens* in intestines; organism in feces of humans, other animals, and in soil	8–22 h, typically 10 h	Abdominal pain, diarrhea	Cooked meat, poultry, gravy, sauces, meat-containing soups, refried beans	Stools	Storing cooked foods at room temperature; storing cooked foods in large containers in refrigerators; holding foods at warm (bacterial-incubating) temperatures; preparing foods several hours before serving; inadequate reheating of leftovers

(continued)

Table B (continued)

Incubation (latency) period usually between 18 and 72 h

Bacteria

Illness	Etiologic agent and source	Incubation or latency period[a]	Signs and symptoms[a]	Foods usually involved[b]	Specimen to collect	Factors contributing to foodborne outbreaks[d]
Aeromonas diarrhea	*Aeromonas hydrophila*	1–2 days	Water diarrhea, abdominal pain, nausea, chills, headache	Fish, shellfish, snails, water	Stools	Contamination of foods by sea or surface water
Campylobacteriosis	*Campylobacter jejuni*, C. coli	2–7 days, usually 3–5 days	Abdominal cramps, diarrhea (blood and mucus frequently in stools), malaise, headache, myalgia, fever, anorexia, nausea, vomiting. Sequela: Guillain–Barré syndrome	Raw milk, poultry, beef liver, raw clams, water	Stools, rectal swabs, blood	Drinking raw milk; handling raw poultry; eating raw or rare meat or poultry; inadequate cooking or pasteurization; cross contamination from raw meat
Cholera	*Vibrio cholerae* serogroup O1 classical and El Tor biotypes; serogroup O139	1–5 days, usually 2–3 days	Profuse, watery diarrhea (rice-water stools), vomiting, abdominal pain, rapid dehydration, thirst, collapse, reduced skin turgor, wrinkled fingers, sunken eyes, acidosis	Raw fish, raw shellfish, crustacea; foods washed or prepared with contaminated water; water	Stools, rectal swabs	Obtaining fish and shellfish from sewage-contaminated waters in endemic areas; poor personal hygiene; infected persons touching foods; inadequate cooking; using contaminated water to wash or freshen foods; improper sewage disposal; using night soil as fertilizer
Cholera-like vibrio gastroenteritis	Non O-1/O139 *V. cholerae* and related spp. (e.g., *V. mimicus*, *V. fluvialus*, *V. hollisae*)	1–5 days	Watery diarrhea (varies from loose stools to cholera-like diarrhea)	Shellfish, fish	Stools, rectal swabs	Obtaining fish and shellfish from sewage-contaminated waters; inadequate cooking; cross contamination

Disease	Etiologic agent	Incubation period	Symptoms	Associated foods	Specimens	Factors contributing
Entero-hemorrhagic or shiga toxin producing *Escherichia coli* diarrhea	*E. coli* O157:H7, O26, O111, O115, O113, other serotypes non-O157 STEC 1026, O45, O103, O111, O121, O145	1–10 days, typically 2–5 days	Watery diarrhea, followed by bloody diarrhea; severe abdominal pain; blood in urine. Sequela: hemolytic uremic syndrome	Undercooked Hamburgers, raw milk, roast beef, sausages, unpasteurized apple juice/cider, sprouts, lettuce, spinach, water	Stools, rectal swabs	Ground beef made from meat from infected cattle; ingesting raw meat or milk; inadequate cooking; cross contamination; infected persons touching ready-to-eat foods; inadequately drying and fermenting meats
Enteroinvasive *Escherichia coli* diarrhea	Enteroinvasive-*E. coli* strains	½–3 days	Severe abdominal cramps, fever, watery diarrhea (blood and mucus usually present), tenesmus, malaise	Salads and other foods that are not subsequently heated; water	Stools, rectal swabs	Inadequate cooking; infected persons touching ready-to-eat foods; not washing hands after defecation; storing cooked foods at room temperature; storing cooked foods in large containers in refrigerators; holding foods at warm (bacterial-incubating) temperatures; preparing foods several hours before serving; inadequate reheating of leftovers
Enterotoxigenic *Escherichia coli* diarrhea	Enterotoxigenic-*E. coli* strains	½–3 days	Profuse watery diarrhea (blood and mucus absent), abdominal pain, prostration, dehydration; low-grade fever, vomiting in a small percentage of cases.	Salads and other foods that are not subsequently heated; soft cheeses, parsley, water	Stools, rectal swabs	Inadequate cooking; infected persons touching ready-to-eat foods; not washing hands after defecation; storing cooked foods at room temperature; storing cooked foods in large containers in refrigerators; holding foods at warm (bacterial-incubating) temperatures; preparing foods several hours before serving; inadequate reheating of leftovers
Plesiomonas enteritis	*Plesiomonas shigelloides*	1–2 days	Diarrhea (blood and mucus in stools), abdominal pain, nausea, chills, fever, headache, vomiting	Water	Stools, rectal swabs	Inadequate treated water

(continued)

Table B (continued)

Illness	Etiologic agent and source	Incubation or latency period[a]	Signs and symptoms[a]	Foods usually involved[b]	Specimen to collect	Factors contributing to foodborne outbreaks[d]
Salmonellosis	*Salmonella* (>2,000 serovars) from feces of infected animals, environment	6–72 h, typically 18–36 h can be longer than 72 h	Abdominal pain, diarrhea, chills, fever, nausea, vomiting, malaise	Poultry, eggs and meat and their products, raw milk and dairy products, other foods contaminated by salmonellae (e.g., sprouts, melons, chocolate, cereal tomatoes, peanut butter, cake batter, spices)	Stools, rectal swabs	Storing cooked foods at room temperature; storing cooked foods in large containers in refrigerators; holding foods (including sliced melons) at warm (bacterial-incubating) temperature; inadequate cooking and reheating; preparing foods several hours before serving; cross contamination; improper cleaning of equipment; obtaining foods from contaminated sources; occasionally infected persons touching ready-to-eat foods
Shigellosis	*Shigella dysenteriae, S. flexneri, S. boydii, S. sonnei*	½–7 days, typically 1–3 days	Abdominal pain, diarrhea (stools may contain blood, pus, and mucus), tenesmus, fever, vomiting	Any ready-to-eat food contaminated by infected person; frequently salads, poi, water	Stools, rectal swabs	Infected person touching ready-to-eat foods, improper refrigeration, inadequate cooking and reheating
Vibrio parahaemolyticus gastroenteritis	*Vibrio parahaemolyticus*	4–96 h, typically 12 h	Abdominal pain, diarrhea, nausea, vomiting, fever, chills, headache	Marine fish, molluscan shellfish, crustacea (raw or recontaminated)	Stools, rectal swabs	Eating raw fin fish and shellfish; inadequate cooking; improper refrigeration; cross contamination; improper cleaning of equipment; using sea water in food preparation or to cool cooked foods
Yersiniosis	*Yersinia enterocolitica*	1–7 days	Abdominal pain (may simulate acute appendicitis); low-grade fever, head-ache, malaise, anorexia, chills, diarrhea, nausea, vomiting	Raw milk, tofu, water chitterlings	Stools, rectal swabs	Inadequate cooking or pasteurization; contamination after cooking; surface or spring water as ingredients or for packing foods; cross contamination

Disease	Organism	Incubation period	Symptoms	Foods	Specimens	Comments
Y. pseudotuberculosis infections	*Yersinia pseudotuberculosis*	1–7 days?	Gastroenteritis is most common, but diarrhea rare. Also, fever and acute abdominal pain due to mesenteric lymphadenitis that mimics appendicitis. Secondary manifestations include erythema nodosum and reactive arthritis. A link to Kawasaki disease in Japan is possible but not confirmed	Homogenized milk, carrots lettuce	Stools	Source in milk unknown, carrots stored over winter were contaminated though shrews, some carrots were spoiled
Viruses						
Astrovirus gastroenteritis	Astroviruses from human feces	1–2 days	Diarrhea, sometimes accompanied by one or more enteric signs or symptoms	Ready-to-eat foods	Stools, acute and convalescent blood	Failure to wash hands after defecation; infected person touching ready-to-eat foods; inadequate cooking or reheating
Incubation Periods from a Few Days to a Few Weeks						
Parasites						
Amoebiasis	*Entamoeba histolytica*	Few days to several months, typically 2–4 weeks	Mild to severe gastroenteritis; abdominal pain, constipation or diarrhea (stools contain blood and mucus), fever, chills, skin ulcers	Raw fruit, vegetable or seafood salads	Stools, blood	Poor personal hygiene; infected persons touching ready-to-eat foods; inadequate cooking and reheating
Anisakiasis	*Anisakis, pseudoterranova*	4–6 weeks	Stomach pain, nausea, vomiting, abdominal pain, diarrhea, fever	Rock fish, herring, cod, salmon, squid, sushi	Stools	Ingestion of raw fish; inadequate cooking

(continued)

Table B (continued)

Illness	Etiologic agent and source	Incubation or latency period[a]	Signs and symptoms[a]	Foods usually involved[b]	Specimen to collect	Factors contributing to foodborne outbreaks[d]
Beef tapeworm infection (Taeniasis)	Taenia saginata from flesh of infected cattle	8–14 weeks	Vague discomfort, hunger pains, loss of weight, abdominal pain	Raw or insufficiently-cooked beef	Stools	Lack of or improper meat inspection; inadequate cooking; inadequate sewage disposal; contaminated pastures
Cyclosporosis	Cyclospora cayetanensis	1–11 days, typically 7 days	Prolonged watery diarrhea, weight loss, fatigue, nausea, anorexia, abdominal cramps	Raspberries, lettuce, basil, water, snow peas	Stools	Sewage contaminated irrigation or spraying water suspected; washing fruits with contaminated water; possibly, handling foods that are not subsequently heated
Cryptosporidiosis	Cryptosporidium parvum	1–12 days, usually 7 days	Profuse watery diarrhea, abdominal pain, anorexia, vomiting, low-grade fever	Apple juice/cider, water, raw milk any food touched by infected food handler	Stools, intestinal biopsy	Inadequate sewage or animal waste disposal; contamination by animal manure; contaminated water; inadequate filtration of water, infected person touching ready-to-eat foods
Fish tapeworm infection (Diphyllobothriasis)	Diphyllobothrium latum from flesh of infested fish	5–6 weeks	Vague gastrointestinal discomfort, anemia may occur	Raw or insufficiently-cooked fresh-water fish (perch, pike, turbot, trout, salmon whitefish)	Stools	Inadequate cooking; improper sewage disposal; sewage-contaminated lakes
Giardiasis	Giardia	5–25 days, typically 7–10 days	Diarrhea (pale, greasy, malodorous stools), abdominal pain, bloating, nausea, weakness, vomiting, dehydration, fatigue, weight loss, fever	Salmon, salads, water	Stools	No or inadequate hand washing after defecation; infected persons handling ready-to-eat foods; inadequate sewage disposal; using untreated surface water supplies as ingredient or for processing
Pork tapeworm infection (Taeniasis)	Taenia solium, Taenia saginata from flesh of infected swine	8–14 weeks	Vague discomfort, hunger pains, weight loss	Raw or insufficiently-cooked pork	Stools	Lack of improper meat inspection; inadequate cooking; improper sewage disposal; contaminated pastures

NEUROLOGICAL SYMPTOMS AND SIGNS (VISUAL DISTURBANCES, TINGLING, AND/OR PARALYSIS) OCCUR

Incubation (latency) period usually less than 1 h

	Toxic agent	Symptoms and signs	Incubation	Organism/vehicle	Specimen	Contributing factors
Fungi						
Ibotenic acid group of mushroom poisoning	Ibotenic acid and muscinol in some mushrooms (mushroom strains are different from those cited on pages 100 and 103)	Drowsiness and state of intoxication, confusion, muscular spasms, delirium, visual disturbances	30–60 min	Amanita muscaria, A. pantherina and related species of mushrooms		Eating A. muscaria and related species of mushrooms; eating unknown varieties of mushrooms; mistaking toxic mushrooms for edible varieties; seeking hallucinogenic effects
Muscarine group of mushroom poisoning	Muscarine in some mushrooms (mushroom strains are different from those cited on pages 100 and 103)	Excessive salivation, perspiration, tearing, reduced pressure, irregular pulse, constricted pupils, blurred vision, asthmatic breathing	15 min to few hours	Clitocybe dealbata, C. rivulosa and many species of Inocybe and Boletus mushrooms		Eating muscarine group of mushrooms; eating unknown varieties of mushrooms; mistaking toxic mushrooms for edible mushrooms
Chemicals						
Organophosphate poisoning	Organic phosphorous insecticides (such as parathion, TEPP, diazinon, malathion)	Nausea, vomiting, abdominal cramps, diarrhea, headache, nervousness, blurred vision, chest pain, cyanosis, confusion, twitching, convulsions	Few minutes to few hours	Any accidentally-contaminated food	Blood, urine, fat biopsy	Spraying foods just before harvesting; storing insecticides in same area as foods; mistaking pesticides for dried foods
Carbamate poisoning	Carbamyl (sevin), Temik (aldicarb)	Epigastric pain, vomiting, abnormal salivation, sweating, twitching, fasciculations, contractions of pupils, muscular incoordination	½ h	Watermelons, cucumbers, any accidentally-contaminated food	Blood, urine	Inappropriate application for vine foods; storing insecticides in same area as foods; mistaking pesticides for powdered foods

(continued)

Table B (continued)

Illness	Etiologic agent and source	Incubation or latency period[a]	Signs and symptoms[a]	Foods usually involved[b]	Specimen to collect	Factors contributing to foodborne outbreaks[d]
Dinoflagellates						
Paralytic/neurologic shellfish poisoning (PSP)	Saxitoxin and similar toxins from dinoflagellates *Alexandrium* and *Gymnodinium* species	Few minutes to 30 min	Tingling, burning, numbness around lips and finger tips, giddiness, incoherent speech, difficulty standing, respiratory paralysis	Mussels, clams, scallops	Gastric washing	Harvesting shellfish from waters with high concentrations of *Alexandrium* or *Gymnodinium* species (Red tides)
Amnesic shellfish poisoning (ASP); one outbreak proven to date (Eastern Canada)	Domoic acid from diatom *Pseudonitzschia pungens f. multiseries*	15 minutes to 38 hours (median 5.5 hours)	Nausea, vomiting, abdominal cramps, headache, diarrhea and disorientation and short term memory loss. The most severely affected had seizures, other neurological symptoms and death	Mussels (also found in clams and crabs worldwide)	Gastric washings or vomitus to detect the diatom species	Eating mussels and other shellfish from unregulated waters, especially after a heavy runoff from agricultural land into aquaculture waters
Toxic animals						
Tetrodotoxin (Fugu/Puffer fish) poisoning	Tetrodotoxin from intestines and gonads of puffer-type fish	10 min to 3 h	Tingling sensation of fingers and toes; dizziness, pallor, numbness of mouth and extremities, gastrointestinal symptoms, hemorrhage, desquamation of skin, fixed eyes, twitching, paralysis, cyanosis; fatalities occur	Puffer-type fish		Eating puffer-type fish; failure to effectively remove intestines and gonads from puffer-type fish if they are to be eaten. Source of toxin most likely *Vibrio alginolyticus* and other *Vibrio* species

Plant toxicants

Jimson weed	Tropane alkaloids	Less than 1 h	Abnormal thirst, photophobia, distorted sight, difficulty speaking, flushing, delirium, coma, rapid heart beat	Any part of jimson weed; tomatoes grafted to jimson weed stock	Urine	Eating any part of jimson weed or eating tomatoes from tomato plant grafted to jimson weed stock
Water hemlock poisoning	Resin and circutoxin in hemlock root *Cicuta virosa, C. masculate*, and *C. douglasii*	15–60 min	Excessive salivation, nausea, vomiting, stomach pain, frothing at mouth, irregular breathing, convulsions, respiratory paralysis	Root of water hemlock	Urine	Eating water hemlock; mistaking water hemlock root for wild parsnip, sweet potato, or carrot

Chemicals

Incubation (latency) period usually between 1 and 6 h

Chlorinated hydrocarbon poisoning	Chlorinated hydrocarbon insecticides	30 min to 6 h	Nausea, vomiting, paresthesia, dizziness, muscular weakness, anorexia, weight loss, confusion	Any accidentally-contaminated food	Blood, urine, stools, gastric washing	Storing insecticides in same area as food; mistaking pesticides for dried foods

Marine Phytoplankton

Ciguatera poisoning	Ciguatoxin in fatty tissues in head and flesh of tropical marine fish. From marine plankton	3–5 h, sometimes longer	Gastrointestinal symptoms which disappear in a few days; tingling and numbness of mouth and limbs, muscular and joint pain, dizziness, cold–hot sensations, rash, weakness, slow heart-beat, prostration, paralysis; neurological problems may last several days; deaths occur	Numerous varieties of tropical fish, e.g. barracuda, grouper, red snapper, amberjack, goatfish, skipjack, parrotfish		Eating fatty tissues in head or flesh of tropical reef fishes; usually large reef fish are more commonly toxic. (The more toxic regions are in the South Pacific and Indian Oceans and the North Caribbean Sea.)

(continued)

Table B (continued)

Illness	Etiologic agent and source	Incubation or latency period[a]	Signs and symptoms[a]	Foods usually involved[b]	Specimen to collect	Factors contributing to foodborne outbreaks[d]
			Incubation (latency) period usually between 12 and 72 h			
			Bacteria			
Botulism	Neurotoxins A, B, E, and F of *Clostridium botulinum*; spores found in soil, freshwater mud and animals. C. botulinum produces neurotoxins A–G but only A, B, E and F are associated with foodborne illnesses. Spores are found in soil, freshwater mud and intestines of animals; type E is mostly associated with marine animals and marine and estuarine sediments.	2 h to 8 days, typically 18–36 h	Gastrointestinal symptoms may precede neurological symptoms. Vertigo, double or blurred vision, dryness of mouth, difficult swallowing, speaking and breathing; descending muscular weakness, constipation, dilated or fixed pupils, respiratory paralysis; fatalities occur	Canned low-acid foods (usually home canned); smoked fish; cooked potatoes; onions, garlic in oil, frozen pot pies, meat loaf, stew left overnight in ovens without heat; fermented fish eggs, fish, marine mammals, muskrat or beaver tails, seal flippers, uneviscerated fish	Blood, stool, gastric washing	Inadequate heat processing of canned foods and smoked fish; postprocessing contamination; uncontrolled fermentations; improper curing of hams and fish; holding foods at room and warm temperatures
			Incubation (latency) period usually greater than 72 h			
			Fungal toxins, Chemicals			
Aflatoxicosis (note there are several other mycotoxin-related illnesses that can give acute symptoms if ingested in sufficient quantities; most cause long-lasting chronic effects)	Aflatoxin and other mycotoxins from *Aspergillus flavus* and *Aspergillus parasiticus*	Days to months	High-level exposures produces an acute hepatic failure (hemorrhage, edema, alteration in digestion, and absorption, cirrhosis), carcinoma of the liver, coma, and death. Chronic exposure in children leads to stunted growth and delayed development	Maize, sorghum, rice, wheat, peanut, soybean, sunflower, cotton), spices, and tree nuts. Also found in eggs, milk, meat of animals exposed to contaminated feed	Urine, blood serum	Baked products made from grain crops infected by *Aspergillus* following prolonged exposure to a high humidity environment or damage from stressful conditions such as drought

Disease	Etiologic agent	Incubation period	Signs and symptoms	Specimen	Foods involved	Factors contributing
Mercury poisoning	Methyl and ethyl mercury compounds from industrial waste and organic mercury in fungicides	1 week or longer	Numbness, weakness of legs, spastic paralysis, impaired vision, blindness, coma	Urine, blood, hair	Grains treated with mercury-containing fungicide; pork, fish and shellfish exposed to mercury compounds	Fish harvested from water polluted with mercury compounds; feeding animals grains treated with mercury fungicides; eating mercury-treated grains or meat from animals fed such grains
Triorthocresyl phosphate poisoning	Triorthocresyl phosphate used as extracts or as oil substitute	5–21 days, mean 10 days	Gastrointestinal symptoms, leg pain, ungainly high-stepping gait, foot and wrist drop	Biopsy of gastronemius muscle	Cooking oils, extracts and other foods contaminated with triorthocresyl phosphate	Using compounds as food extractant or as cooking or salad oil
GENERALIZED INFECTION SIGNS AND SYMPTOMS (FEVER, CHILLS AND/OR MALAISE) OCCUR						
Incubation period usually between 12 and 72 h						
Bacteria						
Vibrio vulnificus infection	*Vibrio vulnificus*	16 h	Septicemia, fever, chills, malaise, prostration; pre-existing liver disease in cases typical	Blood	Raw oysters and clams	Persons with liver ailments eating raw shellfish
Incubation (latency) period usually greater than 1 week						
Bacteria						
Brucellosis	*Brucella abortus, B. melitensis* and *B. suis* from tissues and milk of infected animals	7–21 days	Fever, chills, sweating, weakness, malaise, headache, muscle and joint pain, loss of weight	Blood	Raw milk, goat cheese made from unpasteurized milk	Failure to pasteurize milk, livestock infected with brucellae
Listeriosis	*Listeria monocytogenes*	3–70 days, usually 4–21 days	Fever, headache, nausea, vomiting, stillbirths, meningitis, encephalitis, sepsis	Blood, stool	Coleslaw, milk, soft cheese, pate, turkey franks, processed meats	Inadequate cooking; failure to properly pasteurize milk; prolonged refrigeration

(continued)

Table B (continued)

Illness	Etiologic agent and source	Incubation or latency period[a]	Signs and symptoms[a]	Foods usually involved[b]	Specimen to collect	Factors contributing to foodborne outbreaks[d]
Typhoid and paratyphoid fevers	*Salmonella* Typhi for typhoid from feces of infected humans; or *Salmonella* Paratyphi	7–28 days, usually 14 days	Continued fever, malaise, headache, cough, nausea, vomiting, anorexia, abdominal pain, chills, rose spots, constipation or bloody diarrhea. Sequela: reactive arthritis	Shellfish; any food contaminated by infected person, raw milk, postprocess-contaminated meat, cheese, watercress, water	Stools, rectal swabs, blood in incubatory and early acute phase, urine in acute phase	Infected persons touching foods; failure to wash hands after defecation; chronic excreters inadequate cooking; improper refrigeration; improper sewage disposal; obtaining foods from unsafe sources; harvesting shellfish from sewage–contaminated waters
Tuberculosis	*Mycobacterium bovis*	Long periods, may be months to years to first symptoms	Prolonged illness with fever, cough, night-sweats, weight loss and may occasionally cough up blood: can be fatal	Unpasteurized milk and cheese made from unpasteurized milk	Blood, sputum	Consumption of unpasteurized milk and milk products, especially in countries where *M. bovis* is prevalent in herds
			Viruses			
Hepatitis A	Hepatitis A virus	15–50 days, usually 25–30	Fever, malaise, lassitude, anorexia, nausea, abdominal pain, jaundice, dark urine, light-colored stools	Raw shellfish, any food contaminated by infected person	Stools, urine, blood	Infected persons touching foods; failure to wash hands after defecation; inadequate cooking; harvesting shellfish from sewage-contaminated waters; improper sewage disposal
Hepatitis E	Hepatitis E virus	15–65 days, usually 35–40	Similar to above (high mortality for pregnant women)	Raw shellfish, any food contaminated by infected person	Stools, urine, blood	Infected persons touching foods; failure to wash hands after defecation; inadequate cooking; harvesting shellfish from sewage-contaminated waters; improper sewage disposal

Disease	Etiologic agent	Incubation period	Symptoms and signs	Foods involved	Specimen	Factors contributing
Angiostrongyliasis (eosinophilic meningoen-cephalitis)	*Angiostrongylus cantonensis* (rat lung worm) from rodent feces and soil	14–16 days	Gastroenteritis, headache, stiff neck and back, low-grade fever	Raw crabs, slugs, prawns, shrimp, snails	Blood	Ingesting raw food, inadequate cooking
Toxoplasmosis	*Toxoplasma gondii* from tissue and animal	10–13 days	Fever, headache, myalgia, rash	Raw or insufficiently-cooked beef, lamb, venison, domestic pork	Biopsy of lymph nodes, blood	Ingesting raw meat, inadequate cooking
Trichinosis	*Trichinella spiralis* (roundworm) from flesh of infected wild boar, horse, swine, bear, walrus	4–28 days, mean 9 days	Gastroenteritis, fever, edema about eyes, muscular pain, chills, prostration, labored breathing	Pork, bear meat, walrus flesh; cross contaminated ground beef and lamb, often in grinders, contaminated horse meat	Blood, muscle biopsy, skin test	Eating raw or inadequately cooked pork, boar, bear or horse meat; inadequate cooking or heat processing; feeding uncooked or inadequately heat-processed garbage to swine; failure to clean grinders between grinding pork and other meats

ALLERGIC-TYPE SYMPTOMS AND SIGNS (FACIAL FLUSHING AND/OR ITCHING) OCCUR

Incubation (latency) period usually less than 1 h

Bacterial (and animal) agents

Disease	Etiologic agent	Incubation period	Symptoms and signs	Foods involved	Specimen	Factors contributing
Histamine poisoning (scombroid poisoning)	Histamine-like substance produced by *Morganella morganii* and probably *Proteus* species.	Few minutes to 1 h	Headache, dizziness, nausea, vomiting, peppery taste, burning throat, facial swelling and flushing, stomach pain, diarrhea, itching skin	Tuna, mackerel, Pacific dolphin (mahi mahi), bluefish, cheese		Inadequate cooling; improper refrigeration of fish; improper curing of cheese

(continued)

Table B (continued)

Illness	Etiologic agent and source	Incubation or latency period[a]	Signs and symptoms[a]	Foods usually involved[b]	Specimen to collect	Factors contributing to foodborne outbreaks[d]
			Chemicals			
Monosodium glutamate poisoning	Excessive amounts of monosodium glutamate (MSG)	Few minutes to 1 h	Burning sensation in back of neck, forearms, chest; feeling of tightness in chest, tingling, flushing, dizziness, headache, nausea	Foods seasoned with MSG		Using excessive amounts of MSG as flavor intensifier. Only certain individuals are sensitive to MSG
Nicotinic acid (niacin) poisoning	Vitamin, sodium nicotinate used as color preservative	Few minutes to 1 h	Flushing, sensation of warmth, itching, abdominal pain, puffing of face and knees	Meat or other food in which sodium nicotinate has been added, including baby food and baked goods		Using sodium nicotinate as color preservative, improper mixing

[a]Symptoms and incubation periods will vary with the individual and group exposed because of resistance, age and nutritional status of individuals, number of organisms or concentration of poison ingested, amount of food eaten, and pathogenicity and virulence of strain of microorganism or toxicity of chemical involved. Several of the illnesses exhibit additional symptoms and have incubation periods that are shorter or longer than stated

[b]Collect sample of foods suspected as being the vehicle or contaminated with foodborne pathogens

[c]Carbon monoxide poisoning may simulate this disease. Patients who have been in closed cars with motors running or have been in rooms with improperly vented heaters are subject to exposure to carbon monoxide

[d]For additional information on contributory factors, see Keys A–F

Table C Guidelines for Specimen Collection[††]

Instructions for Collecting Stool Specimens[1]

Instructions	Bacterial	Parasitic[2]	Viral[3]	Chemical
When to collect	During period of active diarrhea (preferably as soon as possible after onset of illness).	Anytime after onset of illness (preferably as soon as possible).	Within 48–72 h after onset of illness.	Soon after onset of illness (preferably within 48 h of exposure to contaminant).
How much to collect	Two rectal swabs or swabs of fresh stool from ten ill persons; samples from ten controls also can be submitted. Whole stool is preferred if nonbacterial stool testing considered	A fresh stool sample from ten ill persons; samples from ten controls also can be submitted. To enhance detection, three stool specimens per patient can be collected >48 h apart.	As much stool sample as possible from ten ill persons (a minimum of 10 mL of stool from each); samples also can be obtained from ten controls.	A fresh urine sample (50 mL) from ten persons; samples from ten controls also can be submitted. Collect vomitus, if vomiting occurs within 12 h of exposure. Collect 5–10 mL whole blood if a toxin/poison is suspected that is not excreted in urine.
Method for collection	For rectal swabs, moisten two swabs in an appropriate transport medium (e.g., Cary-Blair, Stuart, Amies; buffered glycerol–saline is suitable for *E. coli*, *Salmonella*, *Shigella*, and *Y. enterocolitica* but not for *Campylobacter* and *Vibrio*). Insert swab 1–1.5 in. into rectum and gently rotate. Place both swabs into the same tube deep enough that medium covers the cotton tips. Break off top portion of sticks and discard. Alternatively, swab whole stools and put them into Cary-Blair medium.	Collect bulk stool specimen, unmixed with urine, in a clean container. Place a portion of each stool sample into 10% formalin and polyvinyl alcohol preservative (PVA) at a ratio of one part stool to three parts preservative. Mix well. Save portion of the unpreserved stool placed into a leakproof container for antigen or PCR testing.	Place fresh stool specimens (liquid preferable), unmixed with urine, in clean, dry containers, e.g., urine specimen cups.	Collect urine, blood, or vomitus in prescreened containers*. If prescreened containers are not available, submit field blanks with samples[†]. Most analyses from blood require separation of serum from red cells. Cyanide, lead and mercury analyses require whole blood collected in prescreened EDTA tubes. Volatile organic compounds require whole blood collected in a specially prepared gray-top tube.

(continued)

Table C (continued)

Storage of specimens after collection	Refrigerate swabs in transport media at 4°C. When possible, test within 48 h after collection; otherwise, freeze samples at −70°C. Refrigerate whole stool, process it within 2 h after collection. Store portion of each stool specimen frozen at less than −15°C for antigen or PCR testing.	Store specimen in fixative at room temperature, or refrigerate unpreserved specimen at 4°C. A portion of unpreserved stool specimen may be frozen at less than −15°C for antigen or PCR testing.	Immediately refrigerate at 4°C. Store portion of each stool specimen frozen at less than −15°C for antigen or PCR testing.	Immediately refrigerate at 4°C and if possible freeze urine, serum, and vomitus specimens at less than −15°C. Refrigerate whole blood for volatile organic compounds and metals at 4°C.
Transportation	For refrigeration: Follow instructions for viral samples. For frozen samples: Place bagged and sealed samples on dry ice. Mail in insulated box by overnight mail.	For refrigeration: Follow instructions for viral samples. For room-temperature samples: Mail in waterproof container.	Keep refrigerated. Place bagged and sealed specimens on ice or with frozen refrigerant packs in an insulated box. Send by overnight mail. Send frozen specimens on dry ice for antigen or PCR testing.	Immediately refrigerate at 4°C and if possible freeze urine, serum, and vomitus specimens at less than −15°C. Refrigerate whole blood for volatile organic compounds and metals at 4°C. Place double bagged and sealed urine, serum, and vomitus specimens on dry ice. Mail in an insulated box by overnight mail. Ship whole blood in an insulated container with prefrozen ice packs. Avoid placing specimens directly on ice packs.

[1] Label each specimen in a waterproof manner, and put the samples in sealed, waterproof containers (i.e., plastic bags). Batch the collection and send in overnight mail to arrive at the testing laboratory on a weekday during business hours unless other arrangements have been made in advance with the testing laboratory. Contact the testing laboratory before shipping, and give the testing laboratory as much advance notice as possible so that testing can begin as soon as samples arrive. When etiology is unclear and syndrome is nonspecific, all four types of specimens may be appropriate to collect.

[2] For more detailed instructions on how to collect specimens for specific parasites, please go to http://www.CDC.gov and search the web site for key words

[3] For more detailed instructions or how to collect specimens for viral testing, please go to http://www.CDC.gov and search the web site for key words

[*] The containers have been tested for the presence of the chemical of interest prior to use

[†] Unused specimen collection containers that have been brought in to the field and subjected to the same field conditions as the used containers. These containers are then tested for trace amounts of the chemical of interest

[††] from CDC webpage

Table D Guidelines for confirmation of foodborne-disease outbreaks[*]

Etiologic agent	Incubation period	Clinical syndrome	Confirmation
		Bacterial/Chemical/Parasitic/Viral	
Bacterial			
1. *Bacillus cereus*			
a. Vomiting toxin	1–6 h	Vomiting; some patients with diarrhea; fever uncommon	Isolation of organism from stool of two or more ill persons and not from stool of control patients OR Isolation of 10^5 organisms/g from epidemiologically implicated food, provided specimen is properly handled
b. Diarrheal toxin	6–24 h	Diarrhea, abdominal cramps, and vomiting in some patients; fever uncommon	Isolation of organism from stool of two or more ill persons and not from stool of control patients OR Isolation of 10^5 organisms/g from epidemiologically implicated food, provided specimen is properly handled
2. *Brucella*	Several days to several months; usually >30 days	Weakness, fever, headache, sweats, chills, arthralgia, weight loss, splenomegaly	Two or more ill persons and isolation of organism in culture of blood or bone marrow; greater than fourfold increase in standard agglutination titer (SAT) over several wks, or single SAT 1:160 in person who has compatible clinical symptoms and history of exposure
3. *Campylobacter jejuni/coli*	2–10 days; usually 2–5 days	Diarrhea (often bloody), abdominal pain, fever	Isolation of organism from clinical specimens from two or more ill persons OR Isolation of organism from epidemiologically implicated food
4. *Clostridium botulinum*	2 h–8 days; usually 12–48 h	Illness of variable severity; common symptoms are diplopia, blurred vision, and bulbar weakness; paralysis, which is usually descending and bilateral, might progress rapidly	Detection of botulinum toxin in serum, stool, gastric contents, or implicated food OR Isolation of organism from stool or intestine

(continued)

Table D (continued)

Etiologic agent	Incubation period	Clinical syndrome	Confirmation
5. *Clostridium perfringens*	6–24 h	Diarrhea, abdominal cramps; vomiting and fever uncommon	Isolation of 10^6 organisms/g from stool of two or more ill persons, provided specimen is properly handled. OR Demonstration of enterotoxin in the stool of two or more ill persons OR Isolation of 10^5 organisms/g from epidemiologically implicated food, provided specimen is properly handled
6. *Escherichia coli*			
a. Enterohemorrhagic (*E. coli* O157:H7 and others)	1–10 days; usually 3–4 days	Diarrhea (often bloody), abdominal cramps (often severe), little or no fever	Isolation of *E. coli* O157:H7 or other Shiga toxin (verocytotoxin) producing *E. coli* from clinical specimen from two or more ill persons OR Isolation of *E. coli* O157:H7 or other Shiga-like toxin (verocytotoxin) producing *E. coli* from epidemiologically implicated food
b. Enterotoxigenic (ETEC)	6–48 h	Diarrhea, abdominal cramps, nausea; vomiting and fever less common	Isolation of organism of same serotype, demonstrated to produce heat-stable (ST) and/or heat-labile (LT) enterotoxin, from stool of two or more ill persons
c. Enteropathogenic (EPEC)	Variable	Diarrhea, fever, abdominal cramps	Isolation of organism of same enteropathogenic serotype from stool of two or more ill persons
d. Enteroinvasive (EIEC)	Variable	Diarrhea (might be bloody), fever, abdominal cramps	Isolation of same enteroinvasive serotype from stool of two or more ill persons
7. *Listeria monocytogenes*			
a. Invasive disease	2–6 weeks	Meningitis, neonatal sepsis, fever	Isolation of organism from normally sterile site
b. Diarrheal disease	Unknown	Diarrhea, abdominal cramps, fever	Isolation of organism of same serotype from stool of two or more ill persons exposed to food that is epidemiologically implicated or from which organism of same serotype has been isolated

	Incubation period	Symptoms	Confirmation
8. Nontyphoidal *Salmonella*	6 h–10 days; usually 6–48 h	Diarrhea, often with fever and abdominal cramps	Isolation of organism of same serotype from clinical specimens from two or more ill persons OR Isolation of organism from epidemiologically implicated food
9. *Salmonella* Typhi	3–60 days; usually 7–14 days	Fever, anorexia, malaise, headache, and myalgia; sometimes diarrhea or constipation	Isolation of organism from clinical specimens from two or more ill persons OR Isolation of organism from epidemiologically implicated food
10. *Shigella* spp.	12 h–6 days; usually 2–4 days	Diarrhea (often bloody), often accompanied by fever and abdominal cramps	Isolation of organism of same serotype from clinical specimens from two or more ill persons OR Isolation of organism from epidemiologically implicated food
11. *Staphylococcus aureus*	30 min–8 h; usually 2–4 h	Vomiting, diarrhea	Isolation of organism of same phage type from stool or vomitus of two or more ill persons OR Detection of enterotoxin in epidemiologically implicated food OR Isolation of 10^5 organisms/g from epidemiologically implicated food, provided specimen is properly handled
12. *Streptococcus*, group A	1–4 days	Fever, pharyngitis, scarlet fever, upper respiratory infection	Isolation of organism of same M- or T-type from throats of two or more ill persons OR Isolation of organism of same M- or T-type from epidemiologically implicated food

(continued)

Table D (continued)

Etiologic agent	Incubation period	Clinical syndrome	Confirmation
13. *Vibrio cholerae*			
a. O1 or O139	1–5 days	Watery diarrhea, often accompanied by vomiting	Isolation of toxigenic organism from stool or vomitus of two or more ill persons OR Significant rise in vibriocidal, bacterial-agglutinating, or antitoxin antibodies in acute-and early convalescent-phase sera among persons not recently immunized OR Isolation of toxigenic organism from epidemiologically implicated food
b. Non-O1 and non-O139	1–5 days	Watery diarrhea	Isolation of organism of same sterotype from stool of two or more ill persons
14. *Vibrio parahaemolyticus*	4–20 h	Diarrhea	Isolation of Kanagawa-positive organism from stool of two or more ill persons OR Isolation of 10^5 Kanagawa-positive organisms/g from epidemiologically implicated food, provided specimen is properly handled
15. *Yersinia enterocolitica/ Y. pseudotuberculosis*	1–10 days; usually 4–6 days	Gastroenteritis symptoms, abdominal pain (often severe mimicking appendicitis), diarrhea more common with *Y. enterocolitica* infections	Isolation of organism from clinical specimen from two or more ill persons OR Isolation of pathogenic strain of organism from epidemiologically implicated food
Chemical			
1. Marine toxins			
a. Ciguatoxin	1–48 h; usually 2–8 h	Usually gastrointestinal symptoms followed by neurologic symptoms (including paresthesia of lips, tongue, throat, or extremities) and reversal of hot and cold sensation	Demonstration of ciguatoxin in epidemiologically implicated tropical fish OR Clinical syndrome among persons who have eaten a type of tropical fish previously associated with ciguatera fish poisoning (e.g., snapper, grouper, or barracuda)

b. Scombroid toxin (histamine)	1 min–3 h; usually <1 h	Flushing, dizziness, burning of mouth and throat, headache, gastrointestinal symptoms, urticaria, and generalized pruritis	Demonstration of histamine in epidemiologically implicated fish OR Clinical syndrome among persons who have eaten a type of fish previously associated with histamine fish poisoning (e.g., mahi-mahi or fish of order Scomboidei)
c. Paralytic or neurotoxic shellfish poison	30 min–3 h	Paresthesia or lips, mouth or face, and extremities; intestinal symptoms or weakness, including respiratory difficulty	Detection of toxin in epidemiologically implicated food OR Detection of large numbers of shellfish-poisoning-associated species of dinoflagellates in water with which epidemiologically implicated molluscs are gathered
d. Puffer fish, tetrodotoxin	10 min–3 h; usually 10–45 min	Paresthesia of lips, tongue, face, or extremities, often following numbness; loss of proprioception or floating sensations	Demonstration of tetrodotoxin in epidemiologically implicated fish OR Clinical syndrome among persons who have eaten puffer fish
e. Diarrhetic shellfish poison/azaspiracids	minutes – a few hours	Gastrointestinal symptoms (nausea, vomiting, severe diarrhea, abdominal cramps)	Detection of toxin in epidemiologically implicated food Detection of large numbers of shellfish-poisoning associated species of dinoflagellates in water with which epidemiologically implicated molluscs are gathered
2. Heavy metals • Antimony • Cadmium • Copper • Iron • Tin • Zinc	5 min–8 h; usually <1 h	Vomiting, often metallic taste	Demonstration of high concentration of metal in epidemiologically implicated food
3. Monosodium glutamate (MSG)	3 min–2 h; usually <1 h	Burning sensation in chest, neck, abdomen, or extremities; sensation of lightness and pressure over face or heavy feeling in chest	Clinical syndrome among persons who have eaten food containing MSG (e.g., usually 1.5 g MSG)

(continued)

Table D (continued)

Etiologic agent	Incubation period	Clinical syndrome	Confirmation
4. Mushroom toxins			
a. Shorter-acting toxins • Muscimol • Muscarine • Psilogybin • *Coprinus artrementaris* • Ibotenic acid	2 h	Usually vomiting and diarrhea, other symptoms differ with toxin • Confusion, visual disturbance • Salivation, diaphoresis • Hallucinations • Disulfiram-like reaction • Confusion, visual disturbance	Clinical syndrome among persons who have eaten mushroom identified as toxic type OR Demonstration of toxin in epidemiologically implicated mushroom or food containing mushroom
b. Longer-acting toxins (e.g., *Amanita* spp.)	6–24 h	Diarrhea and abdominal cramps for 24 h followed by hepatic and renal failure	Clinical syndrome among persons who have eaten mushroom identified as toxic type OR Demonstration of toxin in epidemiologically implicated mushroom or food containing mushroom
Parasitic			
1. *Cryptosporidium* spp.	2–28 days median: 7 days	Diarrhea, nausea, vomiting; fever	Demonstration of oocysts in stool or in small-bowel biopsy of two or more ill persons OR Demonstration of organism in epidemiologically implicated food
2. *Cyclospora cayetanensis*	1–14 days; median: 7 days	Diarrhea, nausea, anorexia, weight loss, cramps, gas, fatigue, low-grade fever; may be relapsing or protracted	Demonstration of the parasite by microscopy or molecular methods in stool or in intestinal aspirate or biopsy specimens from two or more ill persons OR Demonstration of he parasite in epidemiologically implicated food
3. *Giardia intestinalis*	3–25 days; median: 7 days	Diarrhea, gas, cramps, nausea, fatigue	Demonstration of the parasite in stool or small-bowel biopsy specimen of two or more ill persons

4. *Trichinella* spp.	1–2 days for intestinal phase; 2–4 weeks for systemic phase	Fever, myalgia, periorbital edema, high eosinophil count	Two or more ill persons and positive serologic test or demonstration of larvae in muscle biopsy OR Demonstration of larvae in epidemiologically implicated meat
Viral			
1. Hepatitis A virus	15–50 days; median: 28 days	Jaundice, dark urine, fatigue, anorexia, nausea	Detection of immunoglobulin M antibody to hepatitis A virus (IgM anti-HAV) in serum from two or more persons who consumed epidemiologically implicated food
2. Norovirus (NoV)	12–48 h (median 33 h)	Diarrhea, vomiting, nausea, abdominal cramps, low-grade fever	Detection of viral RNA in at least two bulk stool or vomitus specimens by real-time or conventional reverse transcriptase-polymerase chain reaction (RT-PCR) OR Visualization of viruses (NoV) with characteristic morphology by electron microscopy in at least two or more bulk stool or vomitus specimens OR Two or more stools positive by commercial enzyme immunoassay (EIA)
3. Astrovirus	12–48 h	Diarrhea, vomiting, nausea, abdominal cramps, low-grade fever	Detection of viral RNA in at least two bulk stool or vomitus specimens by real-time or conventional reverse transcriptase-polymerase chain reaction (RT-PCR) OR Visualization of viruses (NoV) with characteristic morphology by electron microscopy in at least two or more bulk stool or vomitus specimens OR Two or more stools positive by commercial enzyme immunoassay (EIA)

*Most etiologic agent descriptions based on information from Centers for Disease Control and Prevention, Atlanta, GA, USA

Table E Guidelines for confirmation of vehicle responsible for foodborne illness

Confirmation status	Criteria
Confirmed vehicle	Isolation of agent from ill persons and from food and laboratory criteria for confirming etiologic agent are as stated in Table D.
	Combination of on-site investigation, statistical evidence and laboratory isolations. (See entries below.)
Presumptive vehicle	Investigation where foods were processed or prepared demonstrates source and mode of contamination and/or survival of etiologic agent in food. Also, desirable to have laboratory isolations of etiolotic agent that causes similar syndrome to that observed during the investigation of food operations and other supportive epidemiologic data. If so, this might provide sufficient evidence for confirmation.
	OR
	p value for food <0.05 when other epidemiologic data support the food hypothesis. Also, desirable to have either laboratory isolations from food or investigation where foods are processed or prepared that demonstrates source and mode of contamination and survival of treatment that supports the hypothesis. If so, this might provide sufficient evidence for confirmation.
	OR
	Odds ratio or relative risk for food greater than 2 and the lower limit of the 95% confidence level is greater than I when other epidemiologic data support a food hypothesis. Also, desirable to have either laboratory isolations from food or investigation in the place foods are processed or prepared that demonstrates source and mode of contamination and survival of treatment that supports the hypothesis. If so, this might provide sufficient evidence for confirmation.

Acknowledgments

The following persons contributed to the development of previous editions of this manual: H.W. Anderson, R.K. Anderson, J. Andrews, J. Archer, F.J. Augulo, K.J. Baker, C.A. Bartleson, F.L. Bryan, R.A. Bryan, H.L. Bryson, M. Cambridge, O.D. Cook, R. Fagan, J. Farber, J.H. Fritz, R. Gilbert, J.J. Guzewich, R.H. Helvig, S.L. Hendricks, L. Jackson, L. Jaykus, V. Lewandowski, M.L. Martin, D. Maxson, H.W. McKinley, D. Morris, J.V. Peavy, E.R. Price, R. Reporter, H.H. Rothe, C. Selman, C.E. Sevey, D. Sharp, P. Sockett, T.E. Sullivan, R.C. Swanson, R. Tauxe, E.C.D. Todd, P.N. Travis, B. Walker, P.C. Wall, I. Weitzman and L. Wisniewski. Their influence on this present edition is acknowledged.

The Committee and Association thank the following persons for their contribution in development of parts of the manual or their peer review of this revision of the 6th edition.

O.D. (Pete) Cook, Office of Regulatory Affairs, Food and Drug Administration, US Department of Health and Human Services, Rockville, MD, USA (Deceased)

Christopher J. Griffith, Food and Consultancy Unit, University of Wales Institute, Cardiff, South Wales, UK

Dean Cliver, Department of Population Health and Reproduction, University of California-Davis, Davis, CA, USA

Tim F. Jones, Tennessee Department of Health, Nashville, TN, USA

Yvonne Salfinger, Florida Department of Agriculture and Consumer Services, Tallahassee, FL, USA

Kristin Delea, US Centers for Disease Control and Prevention, Atlanta, GA, USA

Kirk Smith, Minnesota Department of Health, St Paul, MN, USA

Maja Dobric, University of Georgia, Athens, GA, USA

Ruth Petran, Ecolab, St Paul, MN, USA

Carol Selman, US Centers for Disease Control and Prevention, Atlanta, GA, USA

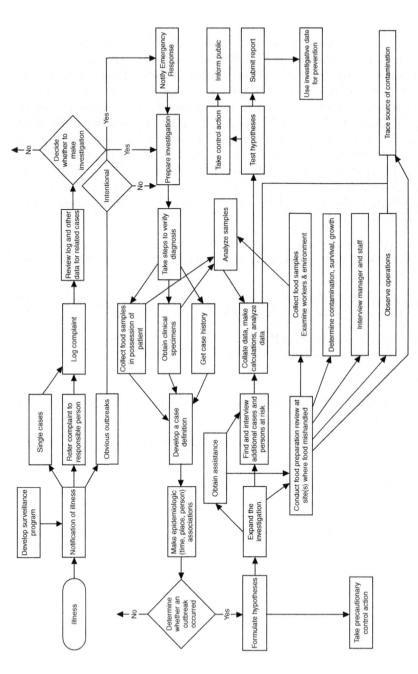

Figure A *Sequence of events in investigating a typical outbreak of foodborne illness.* An intentional food contamination event may or may not be obvious. It can be recognized at any point during the outbreak investigation. If intentional contamination is suspected follow your notification scheme in emergency response plans (this could include law enforcement, emergency management and other government agencies)

FOODBORNE, WATERBORNE, ENTERIC ILLNESS COMPLAINT REPORT
Form A

FOODBORNE, WATERBORNE, ENTERIC ILLNESS COMPLAINT REPORT Form A		Complaint no.*
Complaint received from	Address	Phone Home Work
Person to contact for more information	Address	Phone Home Work e-mail

Complaint
Type of complaint:*   ☐ Illness   ☐ Contaminated/spoiled/adulterated food   ☐ Poor quality drinking water
☐ Poor quality recreational water   ☐ Unsanitary establishment   ☐ Complaint related to media publicity
☐ Disaster   ☐ Other (specify)

Illness: ☐ Yes,[1,2*]   ☐ No   Number ill* _____   Number exposed _____   Time first symptom: Date* _____
Hour _____
Predominant symptoms:*   ☐ Vomiting   ☐ Diarrhea   ☐ Fever   ☐ Neurological   ☐ Skin   ☐ Other (specify)

Physician consulted: ☐ Yes ☐ No If yes, Name	Address	Phone

Hospitalized: ☐ Yes   ☐ No   Emergency Room visit: ☐ Yes   ☐ No
If yes,
Hospital name _____   Address _____
_____   Phone _____
Physician's name _____   Phone _____
Laboratory examination of specimen: Type specimen   Organism/Toxin detected*

Suspect food/water* _____   Source of food/water † _____
Brand identification †   Code/Lot no. †

Suspect meal, event or place:* _____   Date _____   Time _____
Address   Phone

NAME	STATUS	ADDRESS	PHONE
1.	☐ ill ☐ well		
2.	☐ ill ☐ well		
3.	☐ ill ☐ well		
4.	☐ ill ☐ well		

Domestic water source: ☐ Community   ☐ Non-community   ☐ Bottled water
☐ Stream/lake   ☐ Vended   ☐ Well   ☐ Untreated   ☐ Other (specify)

Places and locations where foods eaten past 72 hours, other than home *[3]	Place and locations where water ingested past 2 weeks, other than home *[3]	Place and locations where recreation water contacted past 2 weeks *[3]

History of exposures within past six weeks:*   ☐ Domestic travel (Place) _____
☐ International travel (Place) _____   ☐ Child care   ☐ Contact with ill person outside
household or ill person visited household (indicate name)   ☐ Contact with ill person within household
(indicate name)   ☐ Ill animal _____

Received by	Date of complaint/alert	Time	Disposition

Investigator's name	Comments

[1] If yes, public health professional staff member should obtain information about patient which should be put on Form C.
[2] Ask person to collect vomitus and/or stool in a clean jar, wrap, identify, and refrigerate; hold until health official makes further arrangements.
[3] Ask person to refrigerate all available food eaten during the 72 hours before onset of illness; save or retrieve original containers or packages; sample should be properly identified; hold until health official makes further arrangements. Save any water in refrigerator and trays of ice cubes in freezer; collect was sample from suspect supply in clean jar; put on lid and refrigerate.
* Enter onto complaint log (Form B).
† Enter onto complaint log (Form B) under comments. USE REVERSE SIDE OR ATTACHED SHEET IF MORE SPACE REQUIRED FOR ANY ENTRY

FOODBORNE, WATERBORNE, ENTERIC ILLNESS & COMPLAINT LOG
Form B

COMPLAINT[1]			ILLNESSES			FOOD		WATER[2]			HISTORY OF EXPOSURE[3]	COMMENTS (Specify place, location)
No.	Date	Type	Onset date	No. ill	Predominant symptom/sign	Alleged/Suspected	Where eaten within 72 hrs*	Where ingested within 2 wks*	Where contacted within 2 wks*	Source*		

Legend: [1]Type complaint – I = illness; CF = contaminated/adulterated/spoiledfood; UE = unsanitary food establishment; DW = poor quality drinking water; RW = poor quality recreational water; MP = complaint related to media publicity; D = disaster.

[2]Water source – C = community; NC = non-community; O = other; W = well; B = bottled; S/L = stream/lake; V = vended; U = untreated; O = other

[3]History of exposure (specify by name) – DT = domestic travel (out of town but within country); IT = international travel; CC = child care; CI = contact with ill person outside household or visitor to household; C = contact with ill person within household; AN = exposure to ill animal.

*Enter each place or source on separate line under the complaint number.

CASE HISTORY: CLINICAL DATA
Form C1

Name			Source or place of outbreak, if known	Complaint number	Case identification no.
			Address		Phone: Home Work
Age	Sex	Occupation	Place of work	Ethnic group, special dietary habits, immunocompromised or other pertinent personal or health data

Signs and Symptoms‡ (Check appropriate signs and symptoms and circle those that occurred first)

INTOXICATIONS (Acute and chronic)

☐ Nausea
☐ Vomiting
☐ Anemia
☐ Bloating
☐ Burning sensation (mouth)
☐ Cyanosis
☐ Dehydration
☐ Excessive salivation
☐ Flushing
☐ Foot/wrist drop
☐ Insomnia
☐ Metallic taste
☐ Pallor
☐ Pigmentation
☐ Prostration
☐ Scaling of skin
☐ Soapy/Salty taste
☐ Thirst
☐ Weight loss
☐ White bands on fingernails
☐ Others (specify)

ENTERIC INFECTIONS

*☐ Abdominal cramps
*☐ Diarrhea
  ☐ bloody #
  ☐ greasy
  ☐ mucoid
  ☐ watery
  No./day ____
☐ Chills
☐ Constipation
*☐ Fever ____ °C/°F
☐ Tenesmus

GENERALIZED INFECTIONS

☐ Cough
☐ Edema
☐ Headache
☐ Jaundice
☐ Lack of appetite
☐ Malaise
☐ Muscular aching
☐ Perspiration
☐ Stiff neck joints
☐ Swollen lymph nodes
☐ Weakness
☐ Decrease urine output
☐ Pain in back/kidney

Other INFECTIONS

☐ Ear
☐ Eye
☐ Itching
☐ Mouth
☐ Rash
☐ Skin lesion
☐ Pneumonia
Describe:

NEUROLOGICAL ILLNESSES

☐ Blurred vision
☐ Coma
☐ Delirium
Difficulty in: ☐ speaking
☐ swallowing ☐ breathing
☐ Dizziness
☐ Double vision ☐ Irritability
☐ Disorientation/loss of memory
☐ Hot/cold reversal syndrome
☐ Numbness ☐ Paralysis
Pupils ☐ dilated, ☐ fixed,
  or ☐ constricted
☐ Tingling

Other symptoms	Time of onset Date Hour	Incubation period	Duration of illness	Residual symptoms	Fatel Yes ☐ No ☐

Known allergies	Medications taken for illness		Amount	Dates	Medications/inoculations prior to illness

Physician consulted	Address	Phone	Hospital attended	Address	Phone

Contacts with known cases before illness (names)			Address	Dates of onset	Phone

Cases in household occurring subsequently (names)				Child care exposure (place)	

Type of specimens obtained	Date collected	Specimen number	Laboratory results		Case
1.			Laboratory Method		☐ Confirmed
2.			Laboratory where analysis performed		☐ Presumptive
3.					☐ Suspect

‡Signs and symptoms are listed in columns to suggest classification of the disease; their occurrence is not necessarily limited to the category in which they appear on this form.
*Ask if these symptoms occurred, even if they were not mentioned in the interview.
#Ask whether there was decreased urine output.

CASE HISTORY: FOOD/WATER HISTORY AND COMMON SOURCES
Form C2

☐ Ill   ☐ Well

Date of illness/outbreak[1]	Date _____	Day before illness outbreak Date _____	Two days before illness Date _____
	Breakfast[2]	**Breakfast[2]**	**Breakfast[2]**
	Hour _____	Hour _____	Hour _____
	Place _____	Place _____	Place _____
	Item[3] _____	Item[3] _____	Item[3] _____
	Companions[4] _____	Companions[4] _____	Companions[4] _____
	Lunch[2]	**Lunch[2]**	**Lunch[2]**
	Hour _____	Hour _____	Hour _____
	Place _____	Place _____	Place _____
	Item[3] _____	Item[3] _____	Item[3] _____
	Companions[4] _____	Companions[4] _____	Companions[4] _____
	Dinner[2]	**Dinner[2]**	**Dinner[2]**
	Hour _____	Hour _____	Hour _____
	Place _____	Place _____	Place _____
	Item[3] _____	Item[3] _____	Item[3] _____
	Companions[4] _____	Companions[4] _____	Companions[4] _____
	Non-meal snacks/water ingested[2]	**Non-meal snacks/water ingested[2]**	**Non-meal snacks/water ingested[2]**
	Hour _____	Hour _____	Hour _____
	Place _____	Place _____	Place _____
	Item[3] _____	Item[3] _____	Item[3] _____
	Companions[4] _____	Companions[4] _____	Companions[4] _____

History of ingesting suspect food or water or contact with water from suspect source

Item	Time of eating, drinking or contact		Source	Address
	Date	Hour		

Common events or gatherings	Date	Persons attending[4]	ill	Well	Addresses	Phone

Nonroutine travel past month (international or domestic/locations)	Water supply[5]	Sewage disposal	Pet/Animals (kind and number of each)

Water contacted during recreation or work in last 2 weeks   Unusual water supplies ingested in last 2 weeks

Investigator	Title	Agency	Date

[1] If ill before all meals eaten, complete column for three days before illness and so indicate to obtain 72-hour history.
[2] If water suspected, number of glasses of water, number of cold beverages made with water, number of beverages made with water, number of beverages with ice ingested per day.
[3] Include all foods, ice, water, and other beverages.
[4] Record names of persons eating same meal and whether or not ill.

CASE HISTORIES SUMMARY: CLINICAL DATA
Form D1

Place of outbreak | Dates of outbreak | Complaint number

ID no.	Name of exposed persons whether or not ill	Address	Phone	Sex	Age	Ill	Time of Ingesting food or water or contacting water		Onset of initial Symptom		Incubation period (differences between time of ingesting-contact and onset)	Signs and symptoms								Duration (no. days)	Date			
							Day	Hour	Day	Hour		Nausea	Vomiting	Diarrhea	Abdominal cramps	Fever					Physician seen	Hospitalized	Death

Investigator   Title   Median IP   Remarks

CASE HISTORIES SUMMARY: FOOD/LABORATORY DATA
Form D2

NOTE: Line up with appropriate Identification number of Form D1 (see fold notations above and below)

: (fold mark)

: (fold mark)

Food ingested at suspect meal or event	Water ingested			Laboratory tests	Specific comments or additional information about any ill, not ill persons. Indicate whether ill person is a food worker (if so, name of establishment). (Record all information where space does not permit in other sections, such as additional symptoms, physician, and hospital names.)	Case			ID #
	Glasses water per day	Cold beverages/ water/day	Beverages with ice per day			Suspect	Presumptive	Confirmed	
Suspect food				Confirmed etiology	Remarks				

CLINICAL SPECIMEN COLLECTION REPORT Form E				Complaint no.	Specimen no.
Place of outbreak		Address		Case I.D. no.	Type of specimen
Patient name		Address			Phone

Reason for collecting specimen
□ Victim of outbreak   □ Person at risk but not ill   □ Handler of suspect food or water
□ Suspected carrier   □ Animal   □ Other (specify)

Physician	Address		Phone

Symptoms: □ Nausea   □ Vomiting   □ Diarrhea   □ Fever   □ Other (specify)

Time of ingesting/ contacting suspect food, meal, or water Day Hour	Time of onset Day Hour	Incubation period	Duration of illness	Medications Type	Amount	Dates
Method of collecting specimen			Method of preservation		Method of shipment	

Other Information

Investigator collecting specimen	Title		Agency	Date Hour collected/submitted	
Test requested	Presence/Absence		Count/Titer/ Concentration	Definitive type	

Comments and interpretations

Laboratory analyst	Lab name & location	Date/Hour received	Date started	Date completed	Etiologic agent as determined by analyst

FOOD SAMPLE COLLECTION REPORT Form F		No. of sample units taken	No. of units in lot	Sample no.	Complaint no.
Place collected		Address		Phone	
Person-in-charge	Description of sample or area swabbed	Date/Hour collected		Code/Lot number	
Product name and description	Brand	Type of container			
Name of manufacturer, buyer, seller, importer (as appropriate)	Address	Container size	Production date	Weight/ Size	
Other types of identification	Origin of shipment	Date of shipment		Arrival date	
Bill of lading or contract number		Destination			

Reason for collecting sample: ☐ Food from alleged outbreak   ☐ Ingredient of outbreak food
☐ HACCP analysis   ☐ HACCP verification   ☐ Special survey   ☐ Similar food prepared in similar manner to that involved in outbreak   ☐ Port of entry   ☐ Other (specify) _____

Method of collecting and shipping sample	Collection utensil	Method of sampling ☐ Judgment ☐ Random throughout lot ☐ Random throughout accessible units ☐ Other
	Method of sterilizing container	
Point of operation sample taken	Temperature: Food	Temperature: Storage unit	Time between serving and sampling
Shipped: ☐ Refrigerated ☐ Frozen ☐ Ambient	Carrier	I.D. marks	Cost of sample
Investigator/Sampler	Title	Agency	
Signature of sampler	Signature of representative of party concerned		
Test requested on basis of epidemiologic data	Presence/ Absence	Count/Concentration	Definitive type
Condition of food when received at the laboratory	pH	a_w	Temperature when received
Comments and interpretations by the laboratory			
Laboratory analyst	Laboratory name & location	Date/Hour: Received	Date Started	Date Completed

FLOW PROCESS OF IMPLICATED FOOD Form G		Complaint No.	
Name of Establishment	Person in Charge at Time of Analysis	Title	
Address		Date and time of Analysis	
Product(s) Evaluated (Include brand, code, date received)	Physical Appearance	pH	a_w

Diagram flow process of operation (Insert temperatures and time of processes or delays and make appropriate symbol at exact point in operation.)

Symbols to use in diagram:

+ Likelihood of growth	O Survival likely	△ Inital contamination likely.	△ Contamination by equipment/ utensils
− No growth	× Likelihood of destruction	▽ Worker/person contaminated	
∨ Vegetative cell	s - Bacterial spore		

Investigator		Title	Date

FOOD PROCESSING/PREPARATION HISTORY	Complaint no.
Form H	

Place under investigation _____ Address _____
Owner _____ Plant/Store manager _____ Phone _____
Food being investigated _____ Operation(s) being investigated _____
Date _____ and time _____ of suspect meal Date _____ and time _____ of food preparation, as applicable
Food source/brand _____ Manufacturer _____ Distributor _____
Significant/suspect ingredients _____
Date of delivery _____ Lot code _____ Addresses of source(s) _____

Food characteristics: Temp F/C pH a_w Redox	Upon arrival	Before heating	After heating	During holding	Final product	Time of measurements

SOURCES OF CONTAMINATION (cite or select operations of concern from flow diagram)

	Operation/source	Potential (code)[a]	Observed yes/no	Laboratory confirmed (list pathogen: enter count)
Raw product/Significant ingredient				
Other ingredients of concern				
Condiments/Spices/Additives				
Cross contamination				
(raw to cooked)				
Workers				
Equipment/Utensils				
Cleaning cloths				
Workers:				

Diarrhea or other gastrointestinal sign/symptom or absence from work prior to or during outbreak

Worker's name	Date/time of illness/absence	Illness lab confirmed	Ate suspect food	Job assignment
	/			
	/			
	/			

	Observed	Reported	Name of worker(s)
Touching foods that are not subsequently heated			
Disposable gloves not worn			
Skin infections			
Poor personal hygiene			

Equipment cleaning and sanitizing methods for operation of concern:

Operation _____ Methods _____

Operation _____ Methods _____

Operation _____ Methods _____

Describe other modes of contamination:

[a]Potential codes: 1 — Potential but unlikely; 2 — Potential and sometimes observed or related; 3 — Potential and commonly observed or related; 4 — Potential and almost always observed/found/related

FOOD PROCESSING/PREPARATION HISTORY REPORT (continued)

SURVIVAL	Name, model, location, settings volume, dimensions (as applicable)	Date and time of operation	Time/Temperature exposure records (chart/log data) reported	Time/Temperature exposures during investigation (enter data)
Retorting				
Responsible person(s)	_____	_____/_____		
Equipment used/can size	_____/_____		_____/_____	_____/_____
Food/can	_____		_____/_____	_____/_____
Heat process/Cooking				
Responsible person(s)	_____	_____/_____		
Equipment used	_____		_____/_____	_____/_____
Food	_____		_____/_____	_____/_____
Reheating				
Responsible person(s)	_____	_____/_____		
Equipment used	_____		_____/_____	_____/_____
Food	_____		_____/_____	_____/_____
Other (specify)	_____	_____/_____	_____/_____	_____/_____
Responsible person(s)	_____	_____/_____		
Equipment used	_____		_____/_____	_____/_____
Food	_____		_____/_____	_____/_____
PROLIFERATION				
During refrigerated/frozen transport/delivery/storage				
Responsible person(s)	_____	_____/_____		
Equipment used	_____		_____/_____	_____/_____
Food	_____		_____/_____	_____/_____
After thawing				
Responsible person(s)	_____	_____/_____		
Equipment used	_____		_____/_____	_____/_____
Food	_____		_____/_____	_____/_____
While outdoors				
Responsible person(s)	_____	_____/_____		
Equipment used	_____		_____/_____	_____/_____
Food	_____		_____/_____	_____/_____
While in kitchen				
Responsible person(s)	_____	_____/_____		
Equipment used	_____		_____/_____	_____/_____
Food	_____		_____/_____	_____/_____
During hot/warm holding				
Responsible person(s)	_____	_____/_____		
Equipment used	_____		_____/_____	_____/_____
Food	_____		_____/_____	_____/_____

FOOD PROCESSING/PREPARATION HISTORY REPORT (continued)

PROLIFERATION (continued)	Name, model, location, settings volume, dimensions (as applicable)	Date and time of operation	Time/Temperature exposure records (chart/log data) reported	Time/Temperature exposures during investigation (enter data)
During chilling				
Responsible person(s)	_____	_____/_____		
Equipment used	_____		_____/_____	_____/_____
Food	_____		_____/_____	_____/_____
During cold storage				
Responsible person(s)	_____	_____/_____		
Equipment used	_____		_____/_____	_____/_____
Food	_____		_____/_____	_____/_____
While on cold display				
Responsible person(s)	_____	_____/_____		
Equipment used	_____		_____/_____	_____/_____
Food exposure	_____		_____/_____	_____/_____
Other contributory				
situations (specify)	_____	_____/_____	_____/_____	_____/_____
Responsible person(s)	_____	_____/_____		
Equipment used	_____		_____/_____	_____/_____
Food exposure	_____		_____/_____	_____/_____

Verification of calibration of establishment time-temperature measuring devices. Test using an ice-bath. Record findings below (if temperatures vary from 32°F/0°C, calibrate)

Item _____   Temperature in ice bath _____
Item _____   Temperature in ice bath _____
Item _____   Temperature in ice bath _____
Other calibration procedures

FACTORS CONTRIBUTING TO OUTBREAK (Check all appropriate boxes and describe on back of form)

CONTAMINATION	PROLIFERATION/AMPLIFICATION	SURVIVAL (lack of inactivation)
□ Toxic substance part of tissue □ Poisonous substance intentionally added □ Poisonous or physical substance accidentally/ incidentally added □ Addition of excess quantities of ingredients under these situations are toxic □ Toxic container or pipelines □ Raw product/ingredient contaminated by pathogens from animal or environment □ Prolonged cold storage for several weeks □ Contaminated raw products eaten □ Obtaining foods from polluted sources □ Cross contamination from raw Ingredient of animal origin □ Bare-hand contact by handler/worker/preparer □ Handling by intestinal carrier □ Inadequate cleaning or processing/preparation equipment/utensils □ Storage in contaminated environment □ Other source of contamination (Specify)	□ Allowing foods to remain at room/ warm outdoor temperature _____ for _____ (several) hours □ Slow cooling; depth _____ □ Inadequate cold-holding temperature _____ □ Preparing foods a half day or more before serving; _____ hours □ Insufficient thawing procedure followed by insufficient cooking □ Insufficient time and/or temperature during hot holding _____ time _____ temp □ Insufficient acidification; pH _____ □ Insufficiently low water activity; a_w _____ □ Inadequate thawing of frozen products □ Anaerobic packing/modified atmosphere □ Inadequate fermentation □ Other situations that promoted or allowed microbial growth or toxin production (specify)	□ Insufficient time _____ and/ or temperature _____ during cooking/heat processing □ Insufficient time _____ and/ or temperature during reheating □ Inadequate acidification; pH □ Other process failure (specify)

GRAPH OF TIME-TEMPERATURE MEASUREMENTS[1,2]

Form I

Food_____ Process_____ Complaint No._____

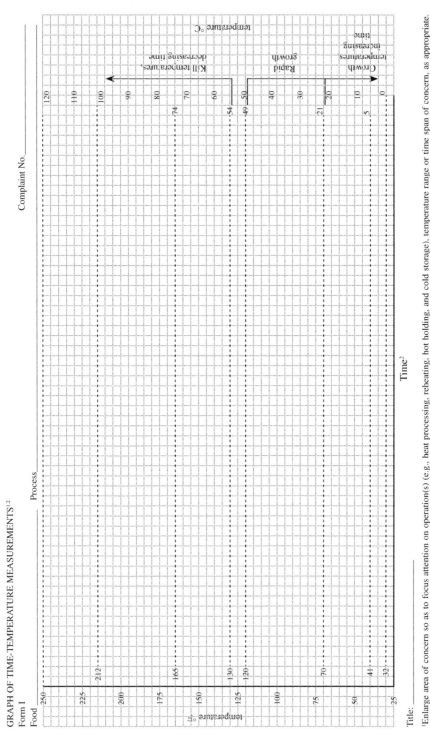

temperature °C

Growth increasing temperatures time

Rapid growth

Kill temperatures, decreasing time

Time[2]

Title:_____

[1]Enlarge area of concern so as to focus attention on operation(s) (e.g., heat processing, reheating, hot holding, and cold storage), temperature range or time span of concern, as appropriate. Fill in time intervals at bottom.

[2]Indicate whether seconds, minutes, hours, days, weeks or other time intervals or time of day.

FOOD TRACEBACK REPORT: PLACES OF SERVICE AND PREPARATION
Form J1

PLACE OF SERVICE INVESTIGATION					Complain/Event no.
Food/ingredient under investigation		Agent			Type/Markers
Place of service[1]		Address			
Owner/Operator		Person interviewed			Phone/Fax
Suspect meal/food product	Date / /	Time	Preparation Date / /		Time
Other meals at which suspect food/ ingredient was served (list meals) _____ _____ _____			Dates served / / / / / /	Known illness _____ _____ _____	No. cases _____ _____ _____
Other dishes/products in which suspect food/ingredient was served/incorporated (list dishes or product) _____ _____ _____			Dates served/ processed / / / / / /	_____ _____ _____	_____ _____ _____
Operations being investigated (e.g., cooking, slicing)			Factors contributing to outbreak		

PLACE OF PREPARATION (If different than place of serving)				
Place prepared/purchased[1]		Address		
Owner/Operator		Person interviewed		Phone/Fax
Label name		Product characteristics (e.g., color, grade, grind size, % fat, size)		
Other meals at which suspect food/ ingredient was served (list meals) _____ _____ _____		Dates served / / / / / /	Known illness _____ _____ _____	No. cases _____ _____ _____
Other dishes/products in which suspect food/ingredient was served/incorporated (list dishes or product) _____ _____ _____		Dates served/ processed / / / / / /	_____ _____ _____	_____ _____ _____
Operations being investigated (e.g., cooking, slicing)		Factors contributing to outbreak at place of service		

PLACE OF PURCHASE OF SUSPECT FOOD OR INGREDIENT				
Supplier[1]		Address		Phone/Fax
Date suspect food/ingredient (lot) received by preparer[2] / /	Quantity received	Lot number	Other product codes/bills of lading numbers	
Manufacturer/Brand		Condition when received (e.g., packaged, loose)		
Product characteristics (e.g., package/container, size/weight/volume, grade)				
Investigator	Title	Agency		Date

[1]Show initials or code used in boxes on flow diagram, Form J3

[2]Attach documentation (e.g., copies of freight bills, air bills, receipts (receiving and sales), signed sworn statements, labels)

FOOD TRACEBACK REPORT: SUPPLIER TO SOURCES OF IMPLICATED[1] FOOD/INGREDIENT Form J2					Complaint/Event no.
SUPPLIER INVESTIGATION[2,3]					Date
Food/ingredient under investigation		Lot code	Agent		Type/Markers
Supplier name		Address	Person interviewed		Phone/Fax
Other shipments of lot of suspect food that could have been present when suspect meal was prepared:					
Brand[2]	Quantity	Lot code	Date received	How used/menu item	Characteristics
1			/ /		
2			/ /		
3			/ /		
4			/ /		
Other consignees to whom the suspect lot was shipped		Address		Phone/Fax	No. persons ill
1					
2					
3					
4					
Factors contributing to contamination, if any			Factors contributing to propagation, if any		
Investigator	Title		Agency		Date
DISTRIBUTOR INVESTIGATION[2,3] (Other middlemen, stops, wholesalers between source and place of service: List in time sequence)					Date
Distributor/whole-sale/shipper name	Address		Person interviewed		Phone/Fax
Shipments received of suspect products		Quantity	Date	Address	Phone/Fax
1			/ /		
2			/ /		
3			/ /		
4			/ /		
Other consignees to whom the suspect lot was shipped		Address		Phone/Fax	No. persons ill
1					
2					
3					
4					
Factors contributing to contamination, if any			Factors contributing to propagation, if any		
Investigator	Title		Agency		Date
SOURCE INVESTIGATION[2,3]					
Name	Location	Person interviewed		Phone	Date(s) of harvest/ production
Factors contributing to contamination, if any			Factors contributing to propagation, if any		
Investigator	Title		Agency		Date

[1]Use additional forms as needed
[2]Attach documentation/identification of contamination or temperature abuse during forward tracing and record on Form H
[3]Laboratory results of samples collected (Attach copy of Form M)

FLOW DIAGRAM OF PRODUCT SOURCE AND DISTRIBUTION	Complaint/event no.		
Form J3	Date		

Food Product	Lot(s) no.	Place of Serving	Number of cases

Illustrate distribution of implicated food/ingredient. Start with place of service, traceback the product flow to its source. Show all suppliers and means of distribution to the source of contamination/survival/propagation or harvester[1]. Also show other consignees that received the contaminated lot(s). Indicate the supplier, distributor, and consignees by a firm code inside a box with arrows showing sequential flow of the food/ingredient. Indicate date of lot movement along side each entry. If additional cases have been identified with serving the implicated food or foods in which the implicated ingredients was or were used, enter these either in or aside the appropriate box.

Investigator	Title	Agency	Phone/Fax

[1]Record complete data (including names of all suppliers, distributors, consignees and the source of the food product; their addresses and phone numbers; and the initials used on this form) on Form J1 and J2

FOOD-SPECIFIC ATTACK RATE TABLE
Form K1

Place of outbreak _____ Complaint no. _____

Food/Beverage	Number of persons who ate				Number of persons who did not eat				Difference In percent	Relative risk	Statistical significance
	a Ill	b Well	a+b Total	Attack rate	c Ill	d Well	c+d Total	Attack rate			

Prepared by: _____ Title _____ Date _____

CASE-CONTROL VEHICLE EXPOSURE TABLE
Form K2

Complaint no.

Place of outbreak

| Food/ Beverage | III | | | | Well | | | | Difference | Odds | Statistical |
	a Ate/drank	c Did not eat	a+c Total	Percent	b Ate/drank	d Did not eat	a+d Total	Percent	In percent	ratio	significance

Prepared by: Title   Date

CALCULATION OF CHI SQUARE TEST, RELATIVE RISK AND ODDS RATIO Form L1					Complaint no.	Place of outbreak	Vehicle
Outbreak table (Step 1)				Expected table (Step 2)			
	III	Well	Total		III	Well	Total
Ate/drank	a	b	a+b	Ate/drank	a_e	b_e	a_e+b_e
Did not eat/drink	c	d	c+d	Did not eat/drink	c_e	d_e	c_e+d_e
Total	a+c	b+d	\<nau\>	Total	a_e+c_e	b_e+d_e	n_e

Explanation	Calculation
Step 1	Step 1
Fill in the outbreak table and calculate the marginal totals (a+b, c+d, a+c, b+d) and the sum of these totals (n) from Form K1 or K2. If any of the marginal totals are less than 10, skip steps 2 through 4 and use Fisher's exact test (Form L2).	*i)†* a + b = _____† *ii)†* c + d = _____† *iii)* a + c = _____ *iv)* b + d = _____ *v)* n = _____
Step 2	Step 2
Fill in the marginal totals in the expected table; copy from those in outbreak table. Calculate the expected frequencies a_e, b_e, c_e, and d_e and fill in the cells of the expected table. If a_e, b_e, c_e, or d_e are less than 5, skip steps 3 and 4 and use Fisher's exact test (Form L2).	*vi)* $a_e =$ $i \times iii/v$ = _____ *vii)* $b_e =$ $i - vi$ = _____ *viii)* $c_e =$ $iii - vi$ = _____ *ix)* $d_e =$ $ii - viii$ = _____
Step 3	Step 3
If *vi*, *vii*, *viii*, and *ix* are greater than 5, calculate the chi-square statistic $$X^2 = \frac{n(|a \times d - c \times d| - c/2)^2}{(a+b)(c+d)(a+c)(b+d)}$$	*x)** a × d = _____* *xi)** b × c = _____* *xii)* x - xi = _____ *xiii)* n/2 = _____ *xiv)* xii - xiii = _____ *xv)* xiv × xiv = _____ *xvi)* xv × n = _____ *xvii)* $i \times ii \times iii \times iv$ = _____ *xviii)* $X^2 = xvi / xvii$ = _____
Step 4	Step 4
Compare X^2 to probability (*p*-value) critical values for the chi square distribution:	*(xviii)* $= X^2 =$ _____ *(xix) p-value =* _____

X^2 - values[1,2]	p-values	
2.71	0.1	Calculate relative risk †RR = $a\sqrt{i} / c\sqrt{ii}$ RR = _____
3.84	0.05	
6.64	0.01	
7.88	0.005	
10.83	0.001	
15.14	0.0001	Calculate odds ratio *OR = x / xi OR = _____
19.51	0.00001	
23.93	0.000001	

[1]X^2 value of 3.84 or greater (*p*<0.05) indicates that there is evidence to suggest a difference between the outbreak table and the expected table, and thus the exposure food/beverage under investigation is related to the observed illness. [2]X^2 value of 7.88 or greater (*p*<0.005) indicates that there is strong evidence to suggest a difference between the outbreak table and the expected table, and thus the exposure food/beverage under investigation is related to the observed illness

CALCULATION OF FISHER'S EXACT TEST Form L2	Complaint number	Place of outbreak	Vehicle

Step 5 (Consider only if steps 3 and 4 are not performed on Form L1)	Formula for calculation $$\frac{(a+b)!\ (c+d)!\ (a+c)!\ (b+d)!}{(n!)\ (a!)\ (b!)\ (c!)\ (d!)}$$

One-tailed test
p1.1 Observed table

Exposure	III	Well	Total	Attack Rate
Ate/drank	a	b	a+b(i)	
Did not eat/ drink	c	d	c+d(ii)	
Total	a+c(iii)	b+d(iv)	n(v)	

vi $\quad p1.1 = \dfrac{(\quad)!(\quad)!(\quad)!(\quad)!}{(\quad!)(\quad!)(\quad!)(\quad!)(\quad!)}$

vii Cancel any possible factorial (!) values
     List individual values from factorials
viii Cancel any possible remaining values
ix   Calculate p1.1 from the remaining values

p1.2 Table

Exposure	III	Well	Total	Attack Rate
Ate/drank	a+1	b−1	a+b(i)	
Did not eat/ drink	c−1	d+1	c+d(ii)	
Total	a+c(iii)	b+d(iv)	n(v)	

vi $\quad p1.2 = \dfrac{(\quad)!(\quad)!(\quad)!(\quad)!}{(\quad!)(\quad!)(\quad!)(\quad!)(\quad!)}$

vii Cancel any possible factorial (!) values
     List individual values from factorials
viii Cancel any possible remaining values
ix   Calculate p1.2 from the remaining values

p1.3 Table

Exposure	III	Well	Total	Attack Rate
Ate/drank	a+2	b−2	a+b(i)	
Did not eat/ drink	c−2	d+2	c+d(ii)	
Total	a+c(iii)	b+d(iv)	n(v)	

vi $\quad p1.3 = \dfrac{(\quad)!(\quad)!(\quad)!(\quad)!}{(\quad!)(\quad!)(\quad!)(\quad!)(\quad!)}$

vii Cancel any possible factorial (!) values
     List individual values from factorials
viii Cancel any possible remaining values
ix   Calculate p1.3 from the remaining values

Etc. continue for all other p-values needed	x $\quad$ p1-value = p1.1 + p1.2 + p1.3 + p1.x for one-tailed test

Interpretation: If the p-value is less than or equal to 0.05, then there is evidence to suggest that the food/beverage under investigation is related to the observed illness; if it is 0.005 or less, there is strong evidence for this relationship.

CALCULATION OF FISHER'S EXACT TEST continued					
Two-tailed test					
p2.1 Table					

Exposure	III	Well	Total	Attack Rate
Ate/drank	a	b=a+b	a+b(i)	
Did not eat/drink	c=a+b	d	c+d(ii)	
Total	a+c(iii)	b+d(iv)	n(v)	

vi $\quad p2.1 = \dfrac{(\quad)!(\quad)!(\quad)!(\quad)!}{(\quad)!(\quad)!(\quad)!(\quad)!(\quad)!}$

vii Cancel any possible factorial (!) values
     List individual values from factorials

$viii$ Cancel any possible remaining values

ix Calculate $p2.1$ from the remaining values

p2.2 Table

Exposure	III	Well	Total	Attack Rate
Ate/drank	a	b=a+b−1	a+b(i)	
Did not eat/drink	c=a+c−1	d	c+d(ii)	
Total	a+c(iii)	b+d(iv)	n(v)	

vi $\quad p2.2 = \dfrac{(\quad)!(\quad)!(\quad)!(\quad)!}{(\quad)!(\quad)!(\quad)!(\quad)!(\quad)!}$

vii Cancel any possible factorial (!) values
     List individual values from factorials

$viii$ Cancel any possible remaining values

ix Calculate $p2.2$ from the remaining values

Etc. for all other p-values

xi $p2\text{-value} = p1 + p2.1 + p2.2$
     $+ p2.x$ for two-tailed test

Interpretation: If the p-value is less than or equal to 0.05, then there is evidence to suggest that the food/beverage under investigation is related to the observed illness; if it is 0.005 or less, there is strong evidence for this relationship

LABORATORY RESULTS SUMMARY
Form M

Complaint no.	Outbreak	Dates of outbreaks

Case/Control (Date from Forms B, C, D, and/or E)

I.D. No.	Case No.	Specimen	Organism/Test result	Marker (serotype, phage pattern, colicin type, toxin type)
___	___	___	___	___
___	___	___	___	___
___	___	___	___	___
___	___	___	___	___
___	___	___	___	___

Food/Environment (Data from Form F)

Sample No.	Sample	Organism/Toxin	Count	Marker
___	___	___	___	___
___	___	___ +	___	___
___	___	___	___	___
___	___	___	___	___

Food worker/handler/preparer. (Data from Forms C, D, and/or E)

I.D. No.	Specimen	Organism	Marker
___	___	___	___
___	___	___	___
___	___	___	___
___	___	___	___

Interpretations and Remarks

Etiologic Agent	Vehicle	Source of Contamination

Prepared by	Title	Date

CONTROL ACTIONS TAKEN AND PREVENTIVE MEASURES RECOMMENDED
Form N

Control actions taken:

Exclusion of infected persons _____ Cases _____ Carriers _____ Contacts of cases _____ (Infected food workers are usually excluded from work when they have signs or symptoms. Occasionally, microbiological tests of specimens are made over a duration of several days or weeks before permission is given to work with foods again. Workers can usually return to work without such testing after they have recovered if there is assurance that they practice good personal hygiene and are effectively supervised.)

Announce the outbreak in the mass media so that the public who purchased the food can be alerted to return it to the place of purchase or other designated location _____; heat or otherwise prepare implicated food safely _____; seek medical consultation/treatment _____; or acquire vaccines; take prophylactic or drugs _____.
(The latter two decisions should be made with consultation from supervisors and medical personnel.)

Seizure of food _____ (Detention [embargo] until tested _____; Removal/destruction _____;
Reprocessed _____; Converted to feed _____; Denatured _____; Buried _____; Other _____)
Reject product _____ Recall of lot _____
(Seizure action may take various forms depending upon the type and degree of contamination and the estimated extent of the contamination and food distributed. Such foods may be held in locked facilities until tested and either released or removed; removed from the premises and reprocessed under supervision, converted to animal feed, denatured, buried, or otherwise destroyed; rejected by processor/preparer or place of use or at port of entry; or recalled (all units of the implicated lots which would be handled as those otherwise removed). The majority of processors/caterers whose foods are under suspicion of being a vehicle voluntarily recall their products despite the associated financial loses and embarrassment.)

Cease preparation of the implicated food until corrections are made _____.
(When a vehicle has been identified, the contributing factors should be corrected before that food is prepared again. The process should be modified so as to avoid or minimize contamination, kill pathogens or inactivate toxins, and prevent or significantly slow growth of pathogenic bacteria so that a recurrence is prevented. Control criteria must be established or followed and the process monitored with sufficient frequency to ensure prevention of the events that lead to the outbreak. The implementation of a hazard analysis critical control point system should be considered for that food and associated operations and perhaps all foods prepared/process/stored in the establishment.)

Closure of premises/establishment _____.
(When imminent risk to health exists if the operations continue, or when contributing factors that cannot be corrected or are continuing, the establishment may be closed. Reopening is considered when the contributing factors are identified and corrected or the operation is brought up to industry standard. Consultation with supervisors is advisable because of legal ramifications, but the prime consideration must be protection of the public health. The majority of operators whose establishment are implicated usually wish to cooperate and may voluntarily offer to close.)

Premises with intentionally contaminated food _____
For food that has been intentionally contaminated, special conditions may apply for disposal, clean-up and re-opening of facilities affected.

Other control actions taken _____ (describe):

Recommendations for prevention of recurrences:

Comments on effectiveness of control actions and preventive measures taken:

Person interviewed_____ Title_____
Investigator _____ Title _____ Date _____

ECONOMIC EVALUATION OF A FOODBORNE DISEASE OUTBREAK Form O				Complaint no.			Disease	
DIRECT COSTS								
	Unit cost	No.	Total cost		Unit cost	No.	Total cost	
Medical				Investigation of illness				
1. Physicians' fees				1. Epidemiological team				
2. Nurses' visits				a. salaries[1]				
3. Hospitalization				b. administration				
a. bed and board				c. other				
b. emergency dept.				2. Laboratory team				
c. acute care				a. salaries[1]				
d. surgery				b. material/equipment				
4. Medication				c. shipping				
5. Ambulance				d. other				
6. Other								
SUBTOTAL				SUBTOTAL				
Loss to food supplier				Loss of productivity				
1. Recall of food				1. Days off work[1]				
2. Storage of food				a. ill person				
3. Destruction/reprocessing				b. enteric pathogen carrier				
4. Laboratory testing, consultant				c. care of ill person				
5. Purchase of new equipment/modification of premises				d. Other personal care				
6. Legal action				2. Workers' compensation payments				
7. Loss of sales				3. Travel to visit sick persons				
8. Increase in insurance premium/bankruptcy				4. Cost of preventive actions				
8. Promotional campaign				5. Other				
9. Other								
SUBTOTAL				SUBTOTAL				
TOTAL DIRECT COSTS								
INDIRECT COSTS								
1. Pain, grief, and suffering[2] =				4. School/study time[2] =				
2. Death[3] =				5. Inability to work at previous occupation[5] =				
3. Leisure time[4] =				6. Other =				
TOTAL INDIRECT COSTS								
TOTAL COSTS =		NUMBER OF CASES				COSTS PER CASE		

[1] Salaries or wages, if not known, can be estimated from the type of occupation reported by ill persons. Daily income can be determined by dividing an annual salary by 365 less days for weekends, holidays and other paid leave; overtime is an extra cost.

[2] Not usually calculated but may be given as a result of a legal settlement.

[3] Calculated on the basis of adjusted willingess-to-pay/human capital estimates (page 67).

[4] Assumed to be equivalent to worth of income.

[5] Calculated on the difference of the incomes before and after illness.

FOODBORNE ILLNESS SUMMARY REPORT Form P	Complaint nos.	Agent and definitive type	Disease
Agency	City	State/Province	
Date of onset of first case	Number ill	Number at risk	Number hospitalized	Fatalities

Symptoms/signs (percentages) Nausea _____ Vomiting _____ Abdominal Cramps _____ Diarrhea _____ Fever _____ Sore/burning mouth/throat _____ Neurological _____ Flushing/itching _____ Other significant (specify) _____	Incubation period Shortest _____ Longest _____ Median _____	Duration Shortest _____ Longest _____ Median _____

Vehicle (Responsible food)	Significant ingredient
Method of processing/preparation	Case definition

PLACE FOOD ACQUIRED (Check one)	SITE OF CONTAMINATION (Check all applicable)	SITE OF SURVIVAL (Check all applicable)	SITE OF PROPAGATION (Check all applicable)	METHOD OF PROCESSING/ PREPARATION (Check all applicable)
☐ Farm	☐ Farm	☐ Farm	☐ Farm	☐ Raw
☐ Aquatic source	☐ Aquatic source	☐ Aquatic source	☐ Aquatic source	☐ Harvest
☐ Woods/lands	☐ Woods/lands	☐ Woods/lands	☐ Woods/lands	☐ Clean/sort/ wash
☐ Food processing	☐ Food processing	☐ Food processing	☐ Food processing	☐ Slaughter/ cut
☐ Bakery	☐ Bakery	☐ Bakery	☐ Bakery	☐ Grind/ blend
☐ Canning	☐ Canning	☐ Canning	☐ Canning	☐ Refrigerated
☐ Egg processing	☐ Egg processing	☐ Egg processing	☐ Egg processing	☐ Frozen
☐ Frozen food	☐ Frozen food	☐ Frozen food	☐ Frozen food	☐ Retorted
☐ Meat	☐ Meat	☐ Meat	☐ Meat	☐ Pasteurized
☐ Poultry	☐ Poultry	☐ Poultry	☐ Poultry	☐ Cooked/ Heated
☐ Seafood	☐ Seafood	☐ Seafood	☐ Seafood	☐ Smoked
☐ Other (specify)	☐ Other (specify)	☐ Other (specify)	☐ Other (specify)	☐ Dried
		☐ Retail outlet		☐ Salted
☐ Retail outlet	☐ Retail outlet	☐ Food service	☐ Retail outlet	☐ Cured
☐ Food service	☐ Food service	☐ Banquet	☐ Food service	☐ Acidified
☐ Banquet	☐ Banquet	☐ Cafeteria	☐ Banquet	☐ Fermented
☐ Cafeteria	☐ Cafeteria	☐ Camp	☐ Cafeteria	☐ Chemically preserved
☐ Camp	☐ Camp	☐ Day	☐ Camp	☐ Vacuum/ anaerobic pack
☐ Day	☐ Day	☐ Military	☐ Day	☐ Mixed/blended
☐ Military	☐ Military	☐ Overnight	☐ Military	☐ Food service
☐ Overnight	☐ Overnight	☐ Recreation	☐ Overnight	☐ Assemble serve
☐ Recreation	☐ Recreation	☐ Catering	☐ Recreation	☐ Cook serve
☐ Catering	☐ Catering	☐ Airline/transport	☐ Catering	☐ Cook hold (ambient)
☐ Airline/Transport	☐ Airline/transport	☐ Banquet	☐ Airline/transport	☐ Cook hold hot
☐ Banquet	☐ Banquet	☐ Central kitchen	☐ Banquet	☐ Cook chill serve
☐ Central kitchen	☐ Central kitchen	☐ Religious/fraternal	☐ Central kitchen	☐ Cook chill serve
☐ Religious/fraternal	☐ Religious/fraternal	☐ Party/social event	☐ Religious/fraternal	☐ Cook chill reheat
☐ Party/social event	☐ Party/Social event	☐ Picnic	☐ Party/ social event	☐ Acidify serve
☐ Picnic	☐ Picnic	☐ Street/office	☐ Picnic	☐ Other (specify)
☐ Street/office	☐ Street/Office	☐ Vending machine	☐ Street/office	
☐ Vending machine	☐ Vending machine	☐ Delicatessen	☐ Vending machine	COMMENTS:
☐ Delicatessen	☐ Delicatessen	☐ Fast food	☐ Delicatessen	
☐ Fast food	☐ Fast food	☐ Ice cream parlor	☐ Fast food	
☐ Ice cream parlor	☐ Ice cream parlor	☐ Industry/office	☐ Ice cream parlor	
☐ Industry/office	☐ Industry/Office	☐ Institution	☐ Industry/office	
☐ Institution	☐ Institution	☐ Hospital	☐ Institution	
☐ Hospital	☐ Hospital	☐ School	☐ Hospital	
☐ School	☐ School	☐ Child care	☐ School	
☐ Child care	☐ Child care	☐ Nursing home	☐ Child care	
☐ Nursing home	☐ Nursing home	☐ Jail/prison	☐ Nursing home	
☐ Jail/prison	☐ Jail/prison	☐ Mental care	☐ Jail/prison	
☐ Mental care	☐ Mental care	☐ Mobile/itinerant	☐ Mental care	
☐ Mobile/itinerant	☐ Mobile/Itinerant	☐ Rooming/tourist home	☐ Mobile/itinerant	
☐ Rooming/tourist home	☐ Rooming/tourist home	☐ Smorgasbord	☐ Rooming/ tourist home	
☐ Smorgasbord	☐ Smorgasbord	☐ Ship	☐ Smorgasbord	
☐ Ship	☐ Ship	☐ Street vending	☐ Ship	
☐ Street vending	☐ Street vending	☐ Table service	☐ Street vending	
☐ Table service	☐ Table service	☐ Take out	☐ Table service	
☐ Take out	☐ Take out	☐ Tavern/bar	☐ Take out	
☐ Tavern/bar	☐ Tavern/Bar	☐ Temporary	☐ Tavern/bar	
☐ Temporary	☐ Temporary	☐ Train	☐ Temporary	
☐ Train	☐ Train	☐ Other (specify)	☐ Train	
☐ Other (specify)	☐ Other (specify)		☐ Other (specify)	
☐ Home	☐ Home	☐ Home	☐ Home	
☐ Residence	☐ Residence	☐ Residence	☐ Residence	
☐ Outdoor (picnic/ beach)	☐ Outdoor (picnic/ beach)	☐ Outdoor (picnic/ beach)	☐ Outdoor (picnic/ beach)	
☐ Potluck gathering	☐ Potluck gathering	☐ Potluck gathering	☐ Potluck gathering	
☐ Private transport	☐ Private transport	☐ Private transport	☐ Private transport	
☐ Other (specify)	☐ Other (specify)	☐ Other (specify)	☐ Other (specify)	

| FACTORS CONTRIBUTING TO OUTBREAK (Check all appropriate) | | |
CONTAMINATION	SURVIVAL (lack of inactivation)	PROLIFERATION/AMPLIFICATION
☐ Toxic substance part of tissue ☐ Poisonous substance intentionally added ☐ Poisonous or physical substance accidentally/incidentally added ☐ Addition of excessive quantities of ingredients that under these situations are toxic ☐ Toxic container or pipelines ☐ Raw product/ingredient contaminated by pathogens from animal or environment ☐ Ingestion of contaminated raw products ☐ Obtaining foods from polluted sources ☐ Cross contamination from raw ingredient of animal origin ☐ Bare-hand contact by handler/worker/ preparer ☐ Handling by intestinal carrier ☐ Inadequate cleaning of processing/ preparation of equipment/utensils ☐ Storage in contaminated environment ☐ Other source of contamination (specify) _____	☐ Insufficient time and/or temperature during cooking/ heat processing ☐ Insufficient time and/or temperature during reheating ☐ Inadequate acidification ☐ Insufficient thawing (followed by insufficient cooking) ☐ Other process failures (specify) _____ _____ _____	☐ Allowing foods to remain at room/ warm outdoor temperature for several hours ☐ Slow cooling ☐ Inadequate cold-holding temperature ☐ Preparing foods a half day or more before serving ☐ Prolonged cold storage for several weeks ☐ Insufficient time and/or temperature during hot holding ☐ Insufficient acidification ☐ Insufficiently low water activity ☐ Inadequate thawing of frozen products ☐ Anaerobic packaging/modified atmosphere ☐ Inadequate fermentation ☐ Other situations that promoted or allowed microbial growth or toxin production (specify) _____

Narrative: (Use additional pages as necessary to give complete story of outbreak)

Attachments with the report:
☐ Case histories Summary (Form D2) ☐ Epidemic curve☐ Laboratory results (Form M) ☐ Food specific attack rate table (Form K1)
☐ Case-control vehicle exposure/dosage (Form K2) ☐ Flow process of implicated food (Form G) ☐ Traceback (Form J)
☐ Food Processing/preparation history and hazard analysis report (Form H) ☐ Graph of time-temperature measurements (Form I) ☐ Control Actions, recommended for prevention (Form N) ☐ Additional narrative☐ Other (specify) _____

_____
_____

Investigator	Reporting agency	Date

Index

CPSIA information can be obtained
at www.ICGtesting.com
Printed in the USA
BVOW07s2126120317

478396BV00020B/97/P

9 781441 983954